EXPERIMENTS IN PLANT PHYSIOLOGY

by

Carol Reiss

Section of Plant Biology,
Cornell University, Ithaca, New York

ILLUSTRATIONS

by

Carol Reiss and Barbara Bernstein

Prentice Hall
Englewood Cliffs, New Jersey 07632

Library of Congress Cataloging-in-Publication Data

Reiss, Carol.
        Experiments in plant physiology/by Carol Reiss; illustrations
    by Carol Reiss and Barbara Bernstein.
            p.   cm.
        Designed to accompany Life with green plant/Arthur W.
    Galston, Peter J. Davies, Ruth L. Slatter. 3rd ed. c1990.
        ISBN 0-13-701285-3
            1. Plant physiology--Experiments.   I. Title.
    QK714.4.R45   1994
    581.1'0724--dc20   93-2239
                                                        CIP

Editorial production: *bookworks*
Acquisitions editor: *David Brake*
Managing editor: *Jeanne Hoeting*
Editor-in-chief: *Tim Bozik*
Prepress buyer: *Paula Massenaro*
Manufacturing buyer: *Lori Bulwin*
Cover photo courtesy of ICI Americas Inc.

 c 1994 by Prentice-Hall, Inc.
A Simon & Schuster Company
Englewood Cliffs, New Jersey 07632

Printed in the United States of America

10 9 8 7 6 5 4 3 2 1

ISBN   0-13-701285-3

Prentice-Hall International (UK) Limited, *London*
Prentice-Hall Australia Pty. Limited, *Sydney*
Prentice-Hall Canada Inc., *Toronto*
Prentice-Hall Hispanoamericana, S.A., *Mexico*
Prentice-Hall of India Private Limited, *New Delhi*
Prentice-Hall of Japan, Inc., *Tokyo*
Simon & Schuster Asia Pte. Ltd., *Singapore*
Editora Prentice-Hall do Brasil, Ltda., *Rio de Janeiro*

CONTENTS

APPENDICES

PREFACE

Plant physiology is one of the experimental sciences; all of our information comes from observation and experimental results. If you only hear the generalizations arising from these experiments--as in most lecture series and textbooks where comprehensive coverage is attempted--your understanding of plant behavior is bound to be a superficial one. Worse yet you may never be able to effectively challenge the lecturer or textbook writer, if even in your own mind. The most important product of your laboratory experience should be some direct contact with the scientific method in action. In taking a laboratory course, you should try to get a feeling for how experiments are made, how to get results and how much (or little) any single experiment can prove. Your laboratory work should be approached in the spirit of scientific inquiry. It does not matter whether the experiment has been done before or not; the work of your own hands will be examined to see if it can yield an answer to a specific question, and as far as **you** are concerned the work has never been done before. In working laboratories it is not at all unusual for a scientist to spend some time trying to repeat a published experiment, both to see if it really works and to see whether an additional facet or interpretation or clue to further productive work might be found which was missed by the original author.

The best experiments will **not** be the ones that simply demonstrate a principle, but the ones that can be framed to give answers to specific analytical questions. Half the art of practicing science lies in knowing how to formulate good questions--the ones that can be answered experimentally with today's tools, and whose answers can throw a little more light on one corner or another of a complex question. I have tried to emphasize the purpose of doing each of these experiments and hope that a question has been asked and answered or a model explored in each. The approach that I have taken is what has traditionally been called "cookbook." I have used this style because I have found that it works best in the classroom; that is, it is most likely to produce the expected results. Although getting "good" results should not be the only goal of working in a student laboratory, teachers recognize that getting "poor" results often leads to discouragement and a belief that "nothing works" in science. In my experience, detailed instructions provide the greatest likelihood of achieving the expected results. To help alleviate the problems usually associated with this approach (i.e., doing without understanding), I have included, wherever possible, the rationale behind each step in the "recipe" with the intention of providing the students with an understanding of the reasoning employed in the design of the experiment. Appendices A through F are intended to clarify and explain in more depth some aspects of technique that are passed over rather quickly in the text.

These experiments are a mix of the classic and the new. Some of the most basic techniques are often the best techniques for teaching experimental science to introductory students. While new and exciting techniques should be included when possible, it is important that these new techniques do not get in the way of the student's understanding of the concepts they are meeting for the first time. I have tried to include both experiments which provide a basic understanding of plant physiology as well as those which introduce students to many of the possible ways to approach problems in biology today. In addition to an introduction to the logic of experimental method, I hope the laboratory will provide an introduction to a few of the many interesting problems remaining in plant physiology, the plant materials suitable for use in their study, and some of the methods that are applicable.

There are far more experiments in this manual than can be completed in a single semester. It is expected that the instructor will select only some of these, based on the emphasis of the particular course and the available equipment. I have tried to make all of the experiments palatable to the instructor (as well as the student) and to provide as much information about the necessary preparations for each experiment as possible. Appendix G lists all of the solutions and materials required for each experiment as well as advice and suggestions for performing the experiment in class. Some experiments contain several parts; in most cases, the parts may stand alone. I encourage the instructor to mix and match these experiments to his or her own taste. The class time needed for completion of the experiments is included in Appendix G to help you in making your choices. Space for data, calculations, tables and graphs is presented at the end of each experiment in this lab manual. These charts and graphs are intended to guide the student in their understanding of the calculations and results. This manual may be used as a workbook or simply as a guide to the structuring of the student's own laboratory notebook. In either case, the questions at the end of the Results sections are intended to explore the student's understanding of the material. They may be assigned individually or used to provide guidance for a written discussion of the results.

Text references are provided for the students so that they may correlate the experiments done in laboratory class with the material learned in lecture class. Additional reading is suggested for those students who wish to pursue a particular interest in the subject matter.

The experiments in this laboratory manual involve the use of a number of poisons and other dangerous chemicals. In general, only very small quantities of these chemicals are suggested for use in this manual; each is selected for what it can tell us about a specific aspect of plant physiology. Use your good common sense and take great care with these substances. Try not to spill them. Do not pipette poison, strong acids or strong alkalis by mouth. Do not smoke, eat or drink in the laboratory at any time. Report all accidents immediately to your instructor. Be sure to read labels carefully and **follow instructions**. Whenever suggested, wear a lab coat, safety glasses and latex gloves. Do not wear contact lenses in the laboratory, as they may hold chemicals near the eye for prolonged periods. Specific information should be given by the instructor for the handling of dangerous substances; take the instructor's advice seriously. The instructor should have chemical safety data sheets available for those substances which might present a hazard to the user. Disposal of chemical agents is a major concern as well. Many reagents may be safely disposed of by flushing down the drain, but a number of chemicals suggested for use by this text must be collected and disposed of in specially marked containers. Be sure to follow the instructor's advice.

Finally, this book, *Experiments in Plant Physiology*, has been designed to accompany the third edtion of *The Life of the Green Plant* by P. J. Davies and A. W. Galston. The sequence of experiments follows the general sequence of materials covered in that text, but could be used in almost any order should another text be employed. Although this manual has been written at a basic, introductory level, many of the experiments which are included are quite sophisticated in concept. Since the experiments can be discussed at a much more in-depth level than presented here, it is hoped that teachers of higher level courses and more advanced students will find these exercises useful as well.

ACKNOWLEDGMENTS

This manual could not have been developed without an awareness of the laboratory manuals which have been published previously by others. Several of the more classic experiments in this text have appeared in different form in other manuals. The ion uptake experiment is based on a similar experiment in *Experiments in Plant Physiology* by C. Ross. The carbon fixation experiment was adapted from *Laboratory Experiments in Plant Physiology* by P. B. Kaufman, J. Labavitch, A. Anderson-Prouty and N. S. Ghosheh. The ascorbic acid experiment was first conceived by T. ap Rees and his original version first appeared in *Laboratory Manual of Cell Biology*, D. Hall and S. Hawkins, eds.

The support of the Plant Biology faculty at Cornell University has been invaluable. I particularly thank Peter J. Davies, Andre T. Jagendorf, and Roger M. Spanswick who originally developed a number of the experiments in this text. In addition, Bob Turgeon and Randy Wayne have contributed advice and assistance. Stephen E. Williams of Lebanon Valley College was the author of the original version of the Plant Movements experiment during his stay at Cornell. Jim Blankenship and Karen Kindle of the Plant Science Center at Cornell provided materials and invaluable guidance in the development of the *Chlamydomonas* transformation experiment. Phil Reid of Smith College is thanked for his guidance, encouragement and help in the design of the tissue printing section. Ron Wolverton provided much needed technical advice and assistance. Barbara Bernstein prepared a number of the drawings and is gratefully acknowledged. I would especially like to thank S. I. Beale of Brown University for providing the use of his computer as well as endless encouragement and advice.

I gratefully thank all of the graduate Teaching Assistants who have taught with me over the last 15 years. They brought their own experience and thought to the in-class experiments and provided suggestions for improving the laboratory manual. In particular, the contributions of David Bruck, Paul Hester, Carol Mapes, Peter Minorsky, John Monroe, and Doug Hamilton (who single-handedly rekindled our ancient GC and developed the ethylene synthesis experiment) are acknowledged. The help provided by John Andersland with formatting the text was invaluable and will be forever appreciated.

I must also recognize with thanks the contribution of the many students who have taken the plant physiology laboratory courses at Cornell over the last 15 years. They have constantly found new ways to challenge my thinking and have taught me that no sentence is ever written clearly enough.

This manual is dedicated to my son Jeremy. I hope that he comes to know all that he can accomplish in his life.

Carol Reiss

EXPERIMENT 1

Determination of the Ascorbic Acid Content of Cabbage

INTRODUCTION

Although you may associate Vitamin C with citrus fruits, ascorbic acid is found in a wide variety of plant tissues. Ascorbic acid is an excellent reducing agent and most likely acts in such a capacity in the plant cell; it is known to be involved in the protection of chlorophyll from singlet oxygen. It is commonly associated with chloroplasts and is present in quantity in green leafy tissues, as found in a head of cabbage.

This experiment has two goals: to determine the amount of ascorbic acid present in fresh cabbage tissue and to determine if boiling the cabbage in water for five minutes destroys the ascorbic acid present in the fresh tissue. Some hints about extracting the ascorbic acid and a method for determining ascorbic acid content are included in the Procedure section. You are asked to design the actual procedure for this experiment yourself, using the information provided. You will then complete the experiment following your own protocol to achieve the goals stated above.

It is hoped that this exercise will introduce you to the problems involved in the design and interpretation of simple experiments and emphasize that experiments with biological systems require at least as much care and attention to detail as experiments in chemistry or physics. The determination of the ascorbic acid content of cabbage will serve to illustrate the general problems involved in quantifying components of the plant cell. Read the discussion that follows (including the Calculation section) and write out your protocol in the space on page 5 before starting the experiment. Be sure to consider what controls are necessary to answer the question, "Does boiling the cabbage destroy the ascorbic acid present in fresh cabbage?"

MATERIALS

Equipment
large knives
balances and weighing paper
400-mL mortars and pestles
sand
Miracloth
150-mm funnels
250-mL and 500-mL graduated cylinders
10-mL pipets
50-mL burets
heat resistant gloves
pH meter
Pasteur pipets and latex bulbs
thermometers

Solutions
ascorbic acid, 4.0 mg/mL (kept cold
    and in the dark)
5% metaphosphoric acid
dichlorophenol-indophenol (DCIP) 0.8 g/L
    NOTE: DCIP may be irritating to the skin;
    use care in handling and dispose of properly

Plant Material
green cabbage (*Brassica oleracea*)

Cleanup
waste container for DCIP

PROCEDURE

The ascorbic acid must first be extracted from the cells by breaking the tissue in a medium suitable for extraction. Cabbage tissue, like many higher plant tissues, can be readily homogenized by grinding in a mortar using a pestle and a small amount of clean sand (to make the process easier).

For an accurate measurement of the ascorbic acid content, the extraction of ascorbic acid must be complete and no ascorbic acid may be lost to degradation. Many plant tissues contain the enzyme ascorbic acid oxidase, which catalyzes the oxidation of ascorbic acid to dehydroascorbic acid (Fig. 1-1). When cells are disrupted by grinding, cell components that are usually separated by membranes (compartmented) mix together. Should this occur, ascorbic acid oxidase may convert the ascorbic acid present in the tissue to dehydroascorbic acid. Grinding the tissue in 5% metaphosphoric acid, which will inactivate the oxidase, will prevent the loss of ascorbic acid to oxidation. Immerse the cabbage in the acid before you begin grinding and, after the tissue is thoroughly ground, filter the homogenate by passing it through Miracloth.

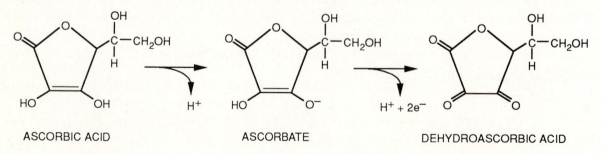

FIGURE 1-1. Ascorbic acid and its oxidation to ascorbate and dehydroascorbic acid.

The hydrogen atoms of the two enol groups of ascorbic acid may be readily oxidized (Fig. 1-1), making ascorbic acid a strong reducing agent. We can take advantage of this property to measure the amount of ascorbic acid present in cabbage. The dye 2,6 dichlorophenol-indophenol (DCIP) is blue in alkali, pink in acid and can be reduced by ascorbic acid to a colorless form (Fig. 1-2). If a drop of the blue dye is added to an acidified extract, the drop will turn pink, then colorless.

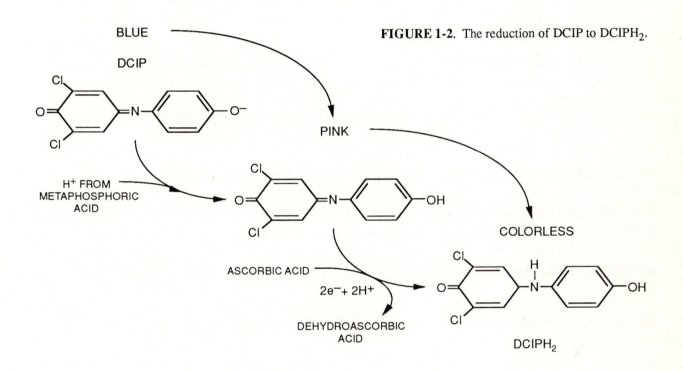

FIGURE 1-2. The reduction of DCIP to DCIPH$_2$.

When all of the ascorbic acid in the extract has been converted to dehydroascorbic acid, no more e$^-$ will be available to reduce a drop of DCIP to the colorless form and the solution will remain pink. Therefore, assuming that ascorbic acid is the only substance present in the cabbage extract that will reduce the dye over the range of pH 1 to 4, the amount of ascorbic acid in an extract can be measured by titration against a dilute solution of dye.

The DCIP solution must first be standardized against a known amount of ascorbic acid. This may be accomplished by titrating the dye into a standard solution containing 1.0 mL of ascorbic acid solution (4.0 mg/mL) and 9 mL of 5% metaphosphoric acid. The end point of the titration will be defined as a pink color that persists through at least 15 seconds of swirling. The amount of ascorbic acid equivalent to 1.0 mL of dye is then calculated. NOTE: Use care in the handling of DCIP. All solutions containing DCIP must be deposited in the DCIP waste container.

Carry out the titrations of your extracts as for the standard and follow the instructions in the Calculations section to determine the ascorbic acid content of fresh tissue and tissue that has been boiled for five minutes. Record your data in the spaces provided on pages 6 and 7. The reported values for the ascorbic acid content of cabbage vary over the range of 20 to 60 mg/100 g fresh weight.

CALCULATIONS

To standardize the dye:

Divide 4.0 mg (the amount of ascorbic acid present in the standard solution) by the number of mL of dye titrated to determine the amount of ascorbic acid equivalent to 1.0 mL of dye:

$$\frac{\text{ascorbic acid (mg)}}{1.0 \text{ mL of dye}} = \frac{4.0 \text{ mg of ascorbic acid}}{\text{dye titrated (mL)}}$$

To determine the average amount of ascorbic acid in an aliquot of extract:

Multiply the average amount (mL) of dye titrated by the amount (mg) of ascorbic acid equivalent to 1.0 mL of dye:

$$\text{mg of ascorbic acid per aliquot} = \text{amount of dye titrated (mL)} \quad \text{x} \quad \frac{\text{ascorbic acid (mg)}}{1.0 \text{ mL of dye}}$$

To determine the amount of ascorbic acid in 100 g of cabbage:

$$\frac{\text{ascorbic acid (mg)}}{100 \text{ g of cabbage tissue}} =$$

$$\text{mg of ascorbic acid per aliquot} \quad \text{x} \quad \frac{\text{total volume of extract (mL)}}{\text{volume of aliquot (mL)}} \quad \text{x} \quad \frac{100}{\text{weight of cabbage (g)}}$$

Determine the amount of ascorbic acid present in your extracts and record the amounts on the chart on page 7. Be sure to consider the number of significant figures that are appropriate in your final answer.

EXPERIMENT 1

Name _____

Date _____

Write out your protocol below. List **all** of the methods used.

RESULTS: Record your data in the spaces provided.

Dye standardization (using 1.0 mL of the ascorbic acid solution and 9 mL of the metaphosphoric acid):

Trial #	Initial Buret Reading	Final Buret Reading	Dye Titrated (mL)
Dye Titrated (average mL):			

amount of ascorbic acid equivalent to 1 mL of dye:

_____

Fresh tissue:

Weight _____

Total volume _____

Aliquot volume _____

Aliquot #	Initial Buret Reading	Final Buret Reading	Dye Titrated (mL)
Dye Titrated (average mL):			

Use the information above to determine the average ascorbic acid content (in mg) in an aliquot:

Then calculate the ascorbic acid content (mg/100 g cabbage tissue) for fresh tissue:

Boiled tissue:

Weight _____

Total volume _____

Aliquot volume _____

Aliquot #	Initial Buret Reading	Final Buret Reading	Dye Titrated (mL)
Dye Titrated (average mL):			

Use the information above to determine the average ascorbic acid content (in mg) in an aliquot:

Then calculate the ascorbic acid content (mg/100 g cabbage tissue) for boiled tissue:

_____:

Weight _____

Total volume _____

Aliquot volume _____

Aliquot #	Initial Buret Reading	Final Buret Reading	Dye Titrated (mL)
Dye Titrated (average mL):			

Use the information above to determine the average ascorbic acid content (in mg) in an aliquot:

Then calculate the ascorbic acid content (mg/100 g cabbage tissue):

Record the amount of ascorbic acid (mg/100 g cabbage tissue) for fresh and boiled tissue. Be sure to include **all** of the ascorbic acid extracted from boiled tissue.

Amount of Ascorbic Acid (mg/100 g F.W.)	
Fresh Tissue	
Boiled Tissue	
Total Ascorbic Acid Extracted from Boiled Tissue	

QUESTIONS

1.   Bearing in mind that a whole cabbage is too cumbersome to grind, what is the ideal tissue sample of cabbage for this experiment? Why? Consider what plant parts a cabbage head contains. Use a drawing to help explain your answer.

2.   What is an aliquot and how is it useful?

3.   If ascorbic acid was lost from the tissue during the boiling procedure, what happened to it? What evidence from your experiment supports this conclusion?

4.   Explain how your choice of boiling procedure affected the total amount of ascorbic acid extracted.

5.   If you were to repeat this experiment, how would you change it to reduce the loss of ascorbic acid during extraction? Be specific.

EXPERIMENT 2

Amylase: Enzyme Assay

INTRODUCTION

Wheat seeds, as well as other cereal seeds, produce the enzyme α-amylase during germination. The substrate for the enzyme α-amylase is stored starch and the end-product is free sugar, which is needed for the growth of the emerging embryo. Extracts from germinating grains like barley and wheat provide a good system for studying enzyme kinetics, since a substantial amount of enzyme is present and the concentration of the substrate (starch) is readily measured when stained with iodine.

Extraction of this enzyme is accomplished by grinding the seeds in buffer to make a puree, then removing unwanted debris through centrifugation. It is essential that the enzyme extract be kept cold at all points during the extraction to prevent the action of proteases which might attack the α-amylase (which is, of course, a protein).

It is not intrinsically difficult to measure enzyme activity. One needs a quantitative measure of either the substrate or the product or both. In this experiment you will provide the starch in a soluble form and add the extracted enzyme. Enzyme activity is sensitive to pH, so you can let the reaction run for a specified time, then stop it by lowering the pH. The undigested starch may be measured spectrophotometrically after staining with iodine.

If samples are taken from a mixture of enzyme and substrate at different times, then evaluated for substrate content, the rate of conversion of substrate into product is determined. A graph of enzyme action (as starch lost) against time is called a **time course**. If enzyme activity is graphed as a function of pH, the optimum pH for enzyme activity may be determined. The sensitivity of α-amylase to pH may be determined by running the reaction for a set time at several pH's. The sensitivity of the extract to high temperature may be tested by boiling the extract before running the reaction.

It should be noted that in this case enzyme activity does not mean "specific activity" (the enzyme activity per mg total protein), but rather the rate of starch digestion as the % starch lost in a given period of time.

MATERIALS

Equipment
ice buckets
mortars and pestles, cold
50-mL centrifuge tubes
refrigerated centrifuge
colorimeter or spectrophotometer
appropriate tubes or cuvettes
plastic bottles, 50 or 100 mL
pipets and pipet aids
test tubes, 25 x 150 mm
test tube racks
boiling water
latex gloves

Solutions
10 mM citric acid-sodium citrate
    buffer at pH 5 (cold)
1 N HCl
iodine solution (5 g KI, 0.36 g
    KIO_3 in 1 liter 2 mM NaOH)
0.05% starch in 0.05 M citric
    acid-sodium citrate buffer at
    pH 3, 4, 5, 6, 7, 8

Plant material
wheat (*Triticum aestivum*) seeds,
germinated for 72 h.

PART I. Preparation of the Enzyme.

PROCEDURE

(1)   Wearing latex gloves, place 50 wheat seeds in a mortar kept cold on ice. Then measure out 40 mL of the cold 10 mM citric acid-sodium citrate buffer and add a small amount to the seeds. Grind the seeds thoroughly, adding more fluid as you go along until about 30 mL have been added.

(2)   Transfer your homogenate to a 50-mL centrifuge tube labeled for your extract, then use the last 10 mL to rinse the mortar, adding this last 10 mL to the centrifuge tube as well. The instructor will collect all the tubes and sediment them in a refrigerated centrifuge for 10 minutes at 15,000 g to remove the starch grains, cell walls, mitochondria and nuclei.

(3)   Pour the supernatant (your extract) into a small plastic bottle and keep the bottle on ice at all times. This enzyme extract will be used for the enzyme assays in Parts II, III and IV.

PART II. The Time Course. The reaction will be run for different time periods.

(1)   Label 16 test tubes (25 x 100 mm) for duplicates of the following time periods: 0, 0.5, 1, 2, 3, 4, 5 and 10 minutes. Add 2 mL of the soluble starch in buffer at pH 5 to each. Make up an additional test tube containing 2 mL of water instead of starch as a background. The background mixture will be used to set the zero on the spectrophotometer or colorimeter to ensure that absorbance by other materials present in the mixture are not interpreted as absorbance by iodine-stained starch.

(2)   Pipet 7 mL of 1 N HCl into the "0 time" treatment and the background tube **before** adding the enzyme. For each of the remaining test tubes, note the starting time and add 1 mL of the enzyme preparation. Let the reaction run for the designated time period, then stop the reaction by adding 7 mL of 1 N HCl and mixing.

(3)   Add 1 mL of the iodine solution to the killed reaction mixture in each tube. The iodine solution develops a blue color when mixed with starch.

(4)   When all of the reactions are complete, transfer the mixtures to colorimeter tubes or cuvettes and measure the color density (or absorbance) of each solution in a colorimeter with a red filter or a spectrophotometer set at 580 nm. Use the background tube to zero the instrument. Record your data in the chart on page 15. Determine the % starch lost (see Calculations) and plot your results on the graph on page 16 before continuing with Part III.

CALCULATIONS

If the background tube was not used to zero your instrument, you must correct your readings by subtracting the value for the background tube from the values of all other tubes.

The percentage of starch lost is obtained from the following procedure:

X represents the amount of starch present at the beginning of the experiment (average of two "0 time" tubes minus the background)
Y represents the amount of starch remaining at the end of the experiment (separately in each of the other tubes minus the background)
X − Y = Z, where Z represents the **amount** of starch lost

Therefore, the starch lost (as %) = $100(Z/X)$

Assume that the colorimeter units (or absorbances) you read are all due to the starch in the substrate solution added, and that there was none lost in the 0-time controls.

With the 0-time controls as 100%, calculate the percent starch lost in each case. If more than 90% of the starch disappeared with any one enzyme preparation, consider the results to be enzyme saturated; comparing results of more than 90% starch lost to each other may be invalid (the reaction may have been limited by substrate).

Plot the % starch lost against time (minutes) on the graph on page 16. Determine the reaction time that will give 75% starch lost and use that as your reaction time for Parts III and IV.

PART III. The pH Curve. The reaction will be run at six different pH's.

(1)   Label 12 test tubes (25 x 100 mm), making duplicates for each of the six pH's: 3, 4, 5, 6, 7 and 8. Add 2 mL of starch solution at the correct pH to the appropriate test tubes.

(2)   From the graph of your time course in Part II, choose a reaction time which will give about 75% starch lost. Note the starting time and add 1 mL of enzyme extract to each test tube. At the end of the reaction time add 7 mL of 1 N HCl and mix.

(3)   Add 1 mL of the iodine solution to the killed reaction mixture in each tube.

(4)   Transfer the mixtures to colorimeter tubes or cuvettes and measure the color density or absorbance of each solution in a colorimeter with a red filter or a spectrophotometer set at 580 nm. Record your data in the chart on page 17.

CALCULATIONS

Determine the % starch lost in each tube (see the Calculations section, Part II). Plot % starch lost against pH before continuing with Part IV (use the graph on page 17).

PART IV. Is the Extract Heat-Labile?

(1)   Prepare 2 test tubes and add 1 mL of the enzyme extract to each. Boil one of the extracts by holding the test tube in boiling water for 2 minutes.

(2)   Add 2 mL of the starch solution at the optimum pH to each tube, mix and let the reaction run for the reaction time that will give about 75% starch lost. Stop the reaction with 7 mL of 1 N HCl.

(3)   Add 1 mL of the iodine solution to each test tube. Transfer the mixtures to colorimeter tubes or cuvettes and read the color density or absorbance on a

colorimeter with a red filter or a spectrophotometer set at 580 nm. Record your data in the chart on page 18.

CALCULATIONS

Determine the % starch lost (as in Part II) for the treatment using boiled extract.

TEXT REFERENCES

Galston, A. W., P. J. Davies and R. L. Satter, *The Life of the Green Plant* (3rd ed.), pp. 29-33. Englewood Cliffs, NJ: Prentice Hall, 1980.

Salisbury, F. B. and C. W. Ross, *Plant Physiology* (4th ed.), pp. 200-205 and pp. 376-380. Belmont, CA: Wadsworth Publishing Co., 1991.

Taiz, L. and E. Zeiger, *Plant Physiology*, pp. 29-39. Redwood City, CA: Benjamin/Cummings Publishing Co., Inc., 1991.

FURTHER READING

Kraut, J., "How do enzymes work?" *Science* (1988) 242:533-540.

EXPERIMENT 2

Name _____

Date _____

RESULTS

PART II.

Record your data in the chart below. Determine the % starch lost at each time period. Show a sample calculation in the space below the chart.

#	Time	Colorimeter Units or A_{580}		Average Units or A_{580}	X – Y = Z	% Starch Lost (100 Z/X)
1 & 2	0			= X		
3 & 4						
5 & 6						
7 & 8						
9 & 10						
11 & 12						
13 & 14						
15 & 16						

Plot the % starch lost against time on the graph on the next page.

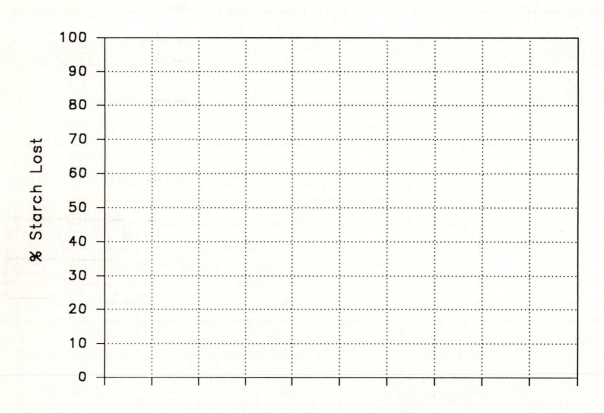

Reaction time giving 75% starch lost _____

PART III.

Record your data for Part III in the chart on the next page. Determine the % starch lost at each pH.

#	pH	Colorimeter Units or A_{580}		Average Units or A_{580}	X – Y = Z	% Starch Lost (100 Z/X)
1 & 2	3					
3 & 4	4					
5 & 6	5					
7 & 8	6					
9 & 10	7					
11 & 12	8					

Plot the % starch lost against pH on the graph below.

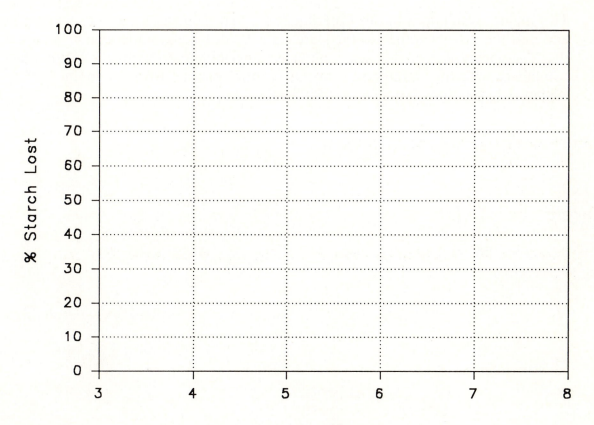

PART IV.

Record your data for Part IV in the chart below.

#	Boiled?	Colorimeter Units or A_{580}	$X - Y = Z$	% Starch Lost (100 Z/X)
1				
2				

QUESTIONS

1. Describe the nature of the curve generated by your time course.

2. What is the optimum pH for amylase action? Does this surprise you? Why or why not?

3. Two major assumptions in this experiment are that the extract contains an enzyme and that the disappearance of the starch is a result of enzymatic action. Couldn't the "disappearance" of the starch be caused by some non-enzymatic factor which complexes with the starch and prevents it from reacting with the iodine? What evidence do you have that an enzyme is involved?

EXPERIMENT 3

Separation, Identification and Quantitation of Plant Pigments

INTRODUCTION

Although most plant leaves appear green to our eyes, several pigments of different color are usually present in the chloroplasts of green leaves. The chlorophylls *a* and *b* provide the green color and absorb the light energy needed for photosynthesis.

Closely associated with the chlorophylls in the chloroplast are another group of pigments, the carotenoids; they are yellow to red in color and likely play a role in the gathering of light energy for photosynthesis. The carotenoids also help to protect the chlorophylls against photooxidation.

Chlorophylls *a* and *b* are tetrapyrrole pigments that contain Mg^{2+} as a centrally chelated metal. Examine Fig. 3-1 and note that, although the structures are otherwise identical, chlorophyll *a* has a methyl group in the position where chlorophyll *b* has a formyl group. The methyl group gives chlorophyll *a* slightly more affinity for non-polar, relatively hydrophobic solvents than chlorophyll *b*. This difference in affinity allows their separation by chromatographic techniques. Pheophytin, which lacks the central Mg^{2+}, is another form of chlorophyll, but it has more affinity for non-polar, hydrophobic solvents. Pheophytin plays a vital role in electron transport, but also may occur as an acid-induced breakdown product of chlorophyll in your spinach extract. Oxalic acid, which is present in quantity in spinach vacuoles, induces pheophytin formation in the ground spinach extract unless special precautions are taken to control the pH. Ask your instructor about the details of the chlorophyll extraction used in your class.

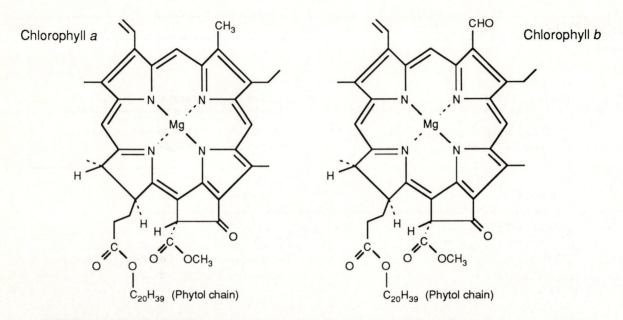

FIGURE 3-1. The chlorophylls *a* and *b*. Note that chlorophyll *a* has a methyl group in the position where chlorophyll *b* has a formyl group.

The carotenoids usually have 40 carbon atoms and are divided into a pure hydrocarbon group, the carotenes, and another group with two additional oxygen atoms, the xanthophylls (Fig. 3-2). The additional oxygens, present as hydroxyl groups at the ends of the molecule, make the xanthophylls more polar and allow them to be separated from the carotenes by chromatographic techniques. In other words, the xanthophylls have less affinity for non-polar solvents than the carotenes.

Using an extract of chloroplast pigment from the leaves of spinach, you will separate the pigments by thin-layer chromatography. You will identify each pigment, then resuspend the chlorophyll *a* and *b* in acetone. The amounts of chlorophylls *a* and *b* will be quantitated and the absorption spectra of several of the pigments determined.

FIGURE 3-2. Two carotenoids.

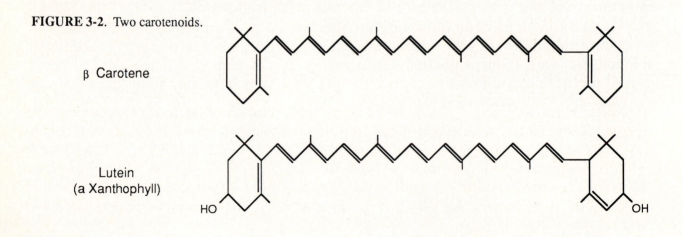

MATERIALS

PART I

Equipment
drying oven, 80°C
silica gel TLC sheets, 2.5 x 7.5 cm
10-μL micropipets
4-oz glass bottles with caps

Solutions
petroleum ether-acetone-chloroform
solution (3:1:1 V/V/V) NOTE:
use in a fume hood and discard in the
proper waste container.

PARTS II and III

Equipment
single-edge razor blades
smooth, white paper squares
Repipets or other pipetting devices
test tubes (12 x 75 mm) and caps
clinical centrifuge
fume hood
spectrophotometer and appropriate tubes or cuvettes
Pasteur pipets and bulbs

Solutions
100% acetone

Clean up
waste container for chlorinated solvents
waste container for acetone

PARTS I, II and III

Plant material
Fresh spinach (*Spinacia oleracea*) tissue (25 g) was homogenized in 100 mL of 90% acetone in a Waring blender. The extract was then filtered through a Büchner funnel and 10 mL of hexane were added. Sufficient distilled water was added (about 20 mL) to bring the final acetone concentration to 70%. The pigment extract was placed in a separatory funnel and the acetone-water phase was removed. The remaining extract was washed twice with distilled water.

PART I. Thin-layer Chromatography (TLC) of Plant Pigments.

Chromatography is a technique that can be used to separate and identify a wide range of compounds. The separation of the components of a mixture is a function of their different affinities for a stationary phase, such as a solid or a liquid, and their differential affinity for a moving phase, such as a liquid or gas. When the stationary phase is solid and the moving phase is liquid, the separation of compounds is governed by their tendency to associate with the mobile (usually hydrophobic) liquid phase or to adsorb onto the solid (usually hydrophilic) surface. The solid phase might be paper, starch, or silica gel. If the solid is applied in a thin layer to a supporting glass or plastic plate, the method is called thin-layer chromatography (TLC). In TLC, the mixture to be separated is first applied as a spot or a line to the solid phase, then the mobile solvent is allowed to pass through the applied compounds along the immobile phase. The compounds will dissolve in and move with the solvent; the movement of the pigments with the solvent along the immobile phase is called the development of the chromatogram. The distance traveled by a particular compound will depend on its affinity for the hydrophobic (mobile) phase versus its affinity for the hydrophilic (solid) phase. This distance is characteristic for a specific set of conditions and may be used to identify the compound. The ratio of the distance traveled by a compound to that of the solvent front is known as the R_f value; unknown compounds may be identified by comparing their R_f's to the R_f's of known standards.

PROCEDURE

(1)   Pour 5 mL of the petroleum ether-acetone-chloroform mixture into a 4-oz bottle, replace the lid and let the chamber equilibrate for five minutes.

(2)   Remove three silica gel plates from the oven and allow them to cool to room temperature. Use forceps and try not to touch the silica gel surface. Mark the **edge** of each plate with a pencil dot about 2 cm above the bottom. With a micropipet, apply 10 μL of concentrated pigment extract in one spot 2 cm above the bottom in the center of the plate (Fig. 3-3). The spot should be made by repeatedly touching the gel with

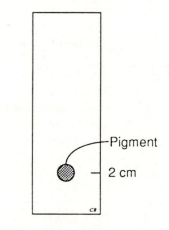

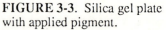

FIGURE 3-3. Silica gel plate with applied pigment.

the micropipet containing 10 μL of the pigment and waiting for the spot to dry between each application. Be careful not to damage the silica gel layer.

(3)   Carefully place **one plate** into the equilibrated chamber and replace the lid. The pigment spot should not yet be in contact with the solvent in the chamber. The solvent will be absorbed by the silica gel and start to move up the plate. When it reaches the pigment spot, the pigments will dissolve in the solvent and move with it along the silica gel. Once a clear separation of the pigments is seen, move to the fume hood, remove the plate and immediately mark the **edge** with a pencil at the solvent front. The solvent will evaporate rapidly, so act quickly. Allow the plate to air dry in the fume hood. Repeat with each of the remaining two plates.

(4)   The pigments visible on the chromatograms are chlorophylls *a* and *b*, carotene, xanthophylls (possibly more than one), and pheophytin. Identify each of the pigments using your knowledge of pigment structure and color.

(5)   Measure the distance traveled by the solvent and the distance traveled by each pigment. The starting point for each measurement is the 2 cm mark. Record your data in the chart on page 27.

(6)   When you are finished, pour the remaining solvent mixture into the waste container provided and leave the bottles in the hood to dry. DO NOT WASH the bottles.

CALCULATIONS

Calculate the R_f values for each pigment. Use the chart on page 27.

$$R_f = \frac{\text{distance traveled by pigment}}{\text{distance traveled by solvent}}$$

Check your results by comparing your R_f values with those listed in Table 3-1. Do these values confirm the identifications you made in step (4)?

TABLE 3-1. R_f's for spinach pigments dissolved in hexane and developed on silica gel plates with pet. ether/acetone/chloroform as the solvent.

Pigment	R_f
Carotene	0.98
Chlorophyll *a*	0.59
Chlorophyll *b*	0.42
Pheophytin	0.81
Xanthophyll 1	0.28
Xanthophyll 2	0.15

PART II. The Quantitation of the Chlorophylls *a* and *b* by Spectrophotometry.

PROCEDURE

(1)   For each of the three plates remove the chlorophyll *a* and *b* bands separately, by scraping the silica gel from the plate with a razor blade onto a square of smooth white paper. Combine all of the chlorophyll *a* scrapings together in one 12 x 75 mm test tube and the chlorophyll *b* scrapings in another. Each test tube will then contain **all** of a particular pigment present in a 30 µL sample of pigment extract.

(2)   Elute the pigments from the silica gel by adding 4 mL of acetone and mixing well. Do not cap the tubes with Parafilm, as it will dissolve in acetone; rubber stoppers or plastic caps may be used.

(3)   Sediment the mixtures for two minutes in a clinical centrifuge to remove the silica gel. Notice the color of the silica gel. Is the silica gel white or has some pigment been retained?

(4)   Use a Pasteur pipet to carefully transfer each eluent to a labeled spectrophotometer tube or cuvette. Be careful not to get fingerprints on the tubes.

(5)   Use acetone as a blank to calibrate the spectrophotometer (see Appendix C). Measure the absorbance of the chlorophyll *a* solution at 663 nm and of the chlorophyll *b* solution at 645 nm. Record your data in the chart on page 27.

As you will determine in Part III, the pigments absorb light to different extents in the different parts of the light spectrum. The most appropriate wavelength to use for measurements of absorbance is at a peak in the absorption spectrum. For chlorophyll *a*, 663 nm is usually selected, while 645 nm is the appropriate choice for chlorophyll *b*.

CALCULATIONS

The higher the concentration of a pigment in a solution, the larger the proportion of light that will be absorbed by the sample at a given wavelength. This relationship is expressed quantitatively by the Beer-Lambert law:

$$A = \epsilon C l \quad \text{or} \quad C = \frac{A}{\epsilon l} \qquad\qquad \text{Eq. (3-1)}$$

A = the absorbance = optical density or OD
ϵ = the millimolar extinction coefficient, which depends on the material and the wavelength and relates the other quantities to each other, in units of L/mmol-cm.
C = the concentration in solution (mmol/L)
l = the light path length (cm)

The concentration of a pigment in solution may be calculated using this equation if the millimolar extinction coefficient is known. Extinction coefficients may be found in reference books such as *Data for Biochemical Research*. For the solvent we use (acetone), ϵ = 75.05 L/mmol-cm for chlorophyll *a* at 663 nm, and ϵ = 47.0 L/mmol-cm for chlorophyll *b* at 645 nm. As an example, Spectronic 20 tubes have an average light path length of 1.17 cm. Therefore:

$$C_{\text{Chl } a} = \frac{A_{663}}{(75.05 \text{ L/mmol-cm})(1.17 \text{ cm})}$$

and

$$C_{\text{Chl } b} = \frac{A_{645}}{(47.0 \text{ L/mmol-cm})(1.17 \text{ cm})}$$

give the **concentration** of these pigments in the acetone solution in units of mmol/L or μmol/mL.

Since we want to determine the **amount** of chlorophyll in a gram of fresh tissue, we must take into account that this sample was dissolved in 4 mL of acetone, and that this was a 30-μL sample out of a 10-mL quantity that contained all the pigments from 25 g of spinach. Therefore:

$$\text{amount of Chl } a = \frac{A_{663}}{(75.05 \text{ mL}/\mu\text{mol-cm})(1.17 \text{ cm})} (4.0 \text{ mL}) \frac{10 \text{ mL}}{0.03 \text{ mL}} \frac{1}{25 \text{ g}}$$

$$= 0.61 \ (A_{663}) \quad \mu\text{mol/g fresh weight}$$

$$\text{amount of Chl } b = \frac{A_{645}}{(47.0 \text{ mL}/\mu\text{mol-cm})(1.17 \text{ cm})} \; (4.0 \text{ mL}) \; \frac{10 \text{ mL}}{0.03 \text{ mL}} \; \frac{1}{25 \text{ g}}$$

$$= 0.97 \; (A_{645}) \quad \mu\text{mol/g fresh weight}$$

Compute these quantities. (NOTE: the light path length for many spectrophotometers is 1 cm rather than 1.17 cm. The above calculations must be adjusted if a spectrophotometer other than the Spectronic 20 is used.) Determine the ratio of the amount of Chl a to Chl b.

PART III. The Determination of the Absorption Spectra for Plant Pigments.

A graph of the absorption of light by a substance, in this case a chloroplast pigment, as a function of the wavelength of visible light is called an **absorption spectrum**. An **action spectrum**, on the other hand, is a graph of the magnitude of a response, for example the rate of photosynthesis, plotted as a function of the wavelength of light. The action spectrum for photosynthesis is shown in Fig. 3-4. You will determine the absorption spectrum for at least one pigment.

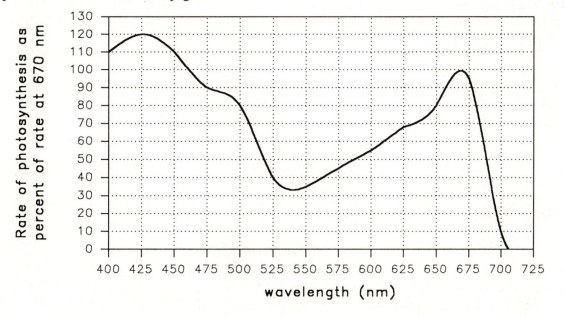

FIGURE 3-4. An action spectrum for photosynthesis.

PROCEDURE

(1) Use one of the following pigments separated by thin-layer chromatography: chlorophyll a, chlorophyll b, carotene or xanthophyll (you may collect all of the carotene or xanthophyll using the method you used to collect the chlorophylls).

(2)   Begin readings at 400 nm and read absorbance values at 25 nm intervals to 700 nm. For chlorophylls *a* and *b*, reduce the intervals to 5 nm between 640 and 670 nm. Recalibrate the instrument at each wavelength with a tube or cuvette containing acetone as a blank. Record your data in the chart on page 28.

(3)   After all of the absorbance values have been obtained in the area of 400-700 nm, plot a graph of absorbance against wavelength on the graph provided (page 28). Examine the absorption spectra completed by other students. Look for major peaks.

NOTE: Absorption spectra may be readily generated using the Diode Array Spectrometer. If one is available and the student has the instructor's approval, it is recommended that all four spectra be completed on the Diode Array Spectrometer. A comparison of the Diode Array Spectrometer with conventional spectrophotometers may be found in Appendix C.

TEXT REFERENCES

Galston, A. W., P. J. Davies and R. L. Satter, *The Life of the Green Plant* (3rd ed.), Chap. 5. Englewood Cliffs, NJ: Prentice Hall, 1980.

Salisbury, F. B. and C. W. Ross, *Plant Physiology* (4th ed.), pp. 209-213. Belmont, CA: Wadsworth Publishing Co., 1991.

Taiz, L. and E. Zeiger, *Plant Physiology*, pp. 181-188. Redwood City, CA: Benjamin/Cummings Publishing Co., Inc., 1991.

FURTHER READING

Clayton, R. K., *Light and Living Matter: A Guide to the Study of Photobiology*, Vol. 1, New York: McGraw-Hill, 1971.

Dawson, R. M. C., D. C. Elliott, W. H. Elliott and K. M. Jones, eds., *Data for Biochemical Research* (3rd ed.). Oxford: Clarendon Press, 1986.

Siefermann-Harms, D., "The light-harvesting and protective functions of carotenoids in photosynthetic membranes," *Physiologia Plantarum* (1987), 69:561-568.

EXPERIMENT 3

Name _____

Date _____

RESULTS

PART I.

Mark the location of each pigment on the diagram below. Measure the distance traveled and calculate the R_f for each pigment.

	Distance Traveled			R_f			Average R_f
	1	2	3	1	2	3	
Solvent Front	___	___	___	xx	xx	xx	xx
	___	___	___	___	___	___	___
	___	___	___	___	___	___	___
	___	___	___	___	___	___	___
	___	___	___	___	___	___	___
	___	___	___	___	___	___	___

PART II.

Determine the amount of chlorophylls a and b in μmol/g fresh weight for your sample.

Pigment	A	Concentration (μmol/mL)	Amount (μmol/g f.w.)	Ratio of Chl a/Chl b

PART III:

Record the values of A for each wavelength in the table below.

Wavelength (nm)	400	425	450	475	500	525	550	575	600
Absorbance (A)									
Wavelength (nm)	625	640	645	650	655	660	665	670	700
Absorbance (A)									

Plot a graph of absorbance against wavelength on the graph below or attach a copy of the spectra generated on the Diode Array Spectrometer.

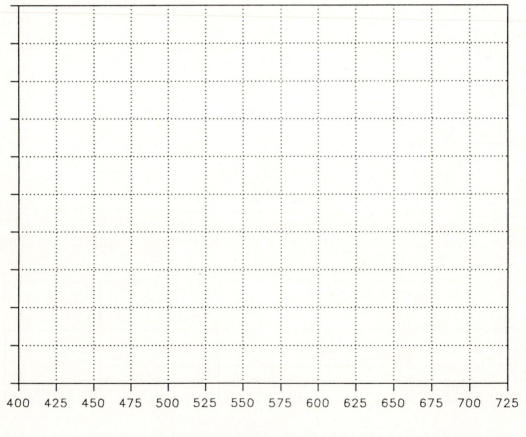

wavelength (nm)

QUESTIONS

1.   What characteristics did you use to identify the pigments that separated on the chromatograms?

2.   The content of chlorophyll *a* is two to three times that of chlorophyll *b* in most green plants. Did your extractions give this ratio? Give two possible reasons why your results might differ.

3.   Which colors of visible light are primarily utilized in photosynthesis?

4.   Compare your absorption spectra with the action spectrum for photosynthesis in Fig. 3-4 or with that in your textbook. Based on the action spectrum for photosynthesis and the absorption spectra obtained by the class what is the primary pigment involved in photosynthesis? Explain.

5.   What are the functions of the other pigments?

EXPERIMENT 4

Whodunit - or - the Influence of Light Intensity on Starch Production in Photosynthesis

INTRODUCTION

The noted botanical detective Chloro Plast had just managed to photograph a notorious plantnapper in the process of pinching some specimens. After the negative was developed, she found that there was insufficient time for her assistant, Calvin, to cycle to the printing shop to purchase photographic paper. Chloro immediately saw the light and realized that she had a supply of plants in her dark room and some iodine in her medicine cabinet. She set to work and shortly had a print of the plant napper.

How was this amazing feat accomplished? Press on and you too can identify the culprit.

MATERIALS

Equipment
large-format, black-and-white negatives
   of a photograph of the culprit
light source
straight sided glass container filled with water
tape
hot plate
tongs
aluminum foil

Solutions
95% ethyl alcohol
Iodine solution

Plant Material
bean (*Phaseolus vulgaris* var.
Redcloud) plants, 2 weeks old,
kept in the dark for 24 h.

PROCEDURE

(1)   In dim room light, place the negative on top of one of the primary leaves (see Appendix F) of a bean plant which has been kept in the dark for 24 hours. Tape the leaf and negative (so that the negative is between the leaf and the light source) onto the side of a glass container filled with water. It is important that the negative make good contact with the leaf (or your print will be blurry), but that a minimum amount of tape is used. The lower surface of the leaf must be exposed to the air to allow for gas exchange. Shine direct light (lamplight or sunlight) through the glass container and onto the leaf for two hours (Fig. 4-1).

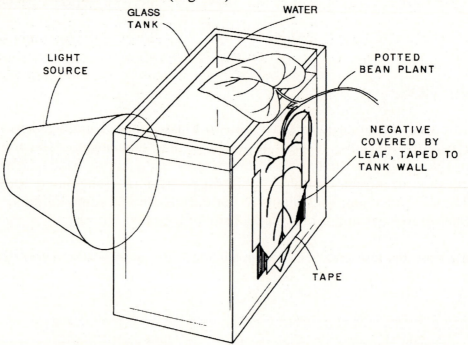

FIGURE 4-1. With the negative flat against the upper surface of the leaf, tape the leaf onto the side of a glass container filled with water.

(2)   Remove the negative from the leaf, cut the leaf from the plant and treat the leaf as directed in step (3).

(3)   Hold the leaf with forceps and dip it in:

   a)   boiling water for 10 seconds to kill the cells.

   b)   hot ethyl alcohol (do this in the hood) for one minute to remove most of the chlorophyll.

   c)   ethyl alcohol for two minutes. The leaf should be almost white.

(4)   Blot the leaf on paper towels.

(5)   Soak the leaf in the iodine solution for one minute or less. You are developing a print, so remove it when the picture is clearly seen.

(6)   Place the leaf in a dish of water containing a square of aluminum foil. Then gently spread the leaf with forceps so that the print is visible. When the leaf is smooth, remove both the foil and leaf together to ensure that the leaf remains flat. Blot the leaf with paper towels and press it in a book.

If the exposure was right you should have a positive print of the negative.

Iodine is a specific stain for starch, producing a black color (actually very dark blue) where starch is present. The development of a print can be explained in terms of the influence of light intensity on the process of photosynthesis, the final product of which is the starch deposited in the chloroplasts.

Your leaves may be dried and mounted below. Note the relationship between the areas of high light transmission through the negative and the areas of starch production.

EXPERIMENT 5

Photosynthesis: Partial Reactions in Cell-Free Preparations

INTRODUCTION

Photosynthesis consists of a complex series of reactions which can be divided into two major phases, carried out in two separate parts of the chloroplast. The reception of light, the electron transport reactions (with water as initial donor – oxygen is given off as the waste product – and $NADP^+$ as the final electron acceptor) and the associated phosphorylation of ADP to ATP all take place in the green, folded lamellar membranes that run through the inside of the chloroplast. Surrounding the lamellar system is the stroma, in turn surrounded by a double outer membrane; the stroma is where soluble enzymes fix carbon dioxide and carry on the metabolism to convert it to carbohydrates, using as a driving force the NADPH and ATP produced in the light reactions. This experiment will focus on the light reactions, the membrane bound steps of photosynthesis called the Hill Reaction.

When chloroplasts are isolated, the stroma and its enzymes are largely lost (unless special precautions are taken). The resulting stripped lamellar system (thylakoid membranes) can still carry on electron transport and photosynthetic phosphorylation if it is provided with an appropriate electron acceptor and the substrates for phosphorylation. You will perform three procedures and monitor rates of the light reactions in thylakoid membranes at three different sites. In Part I you will use DCIP as the final electron acceptor and measure the reduction of this dye by electrons from photosystem I (PSI) using spectrophotometry. In Part II (using ferricyanide as the final electron acceptor), you measure the rate of oxygen evolution at photosystem II (PSII) using an oxygen electrode (see Appendix D). Finally, in Part III, you will measure rates of photophosphorylation in the presence of the electron acceptor phenazine methosulfate.

In Part I, the dye DCIP (the same indophenol dye used in Expt. 1) will be used as the terminal electron acceptor. You should remember (see Fig. 1-2) that it is blue in the oxidized form (DCIP) and colorless in the reduced form ($DCIPH_2$). By measuring the disappearance of blue color (as the dye is reduced by electrons from PSI), you will be able to measure the rate of electron transport.

To measure oxygen evolution in Part II, you will use potassium ferricyanide ($K_3Fe(CN)_6$) as the electron acceptor and observe the change in oxygen tension in solution by means of the oxygen electrode. The rate of oxygen evolution at PSII gives another measure of the rate of electron transport. Ferricyanide is readily reduced to ferrocyanide ($K_4Fe(CN)_6$). The DCIP dye used in Part I is too densely colored for use in the oxygen electrode; the high concentration of dye needed to get measurable amounts of oxygen evolution would badly shade the chloroplast preparation.

Finally, in Part III you will look at photophosphorylation, dependent in this case on a cyclic electron flow. Cyclic electron transport will occur when the electron acceptor used (phenazine methosulfate or PMS) is reduced by electrons from PSI, picking up a H^+ from the stroma. PMSH readily crosses the thylakoid membrane and donates an electron back to PSI, depositing a H^+ inside the thylakoid. The net result is a cycling of electrons around PSI and the buildup of a high internal H^+ concentration allowing for continued photophosphorylation (Fig. 5-1).

Phenazine methosulfate is used for the phosphorylation measurement because: a) ferricyanide and DCIP phosphorylation rates are slower; b) if DCIP were used, not enough phosphorylation would occur before the DCIP was exhausted; c) using more DCIP would shade the chloroplasts too much; d) ferricyanide interferes with the method for measuring inorganic phosphate.

Inorganic phosphate is a substrate of this reaction and can be detected colorimetrically. Phosphate forms a complex with the molybdate ion in the acidic "phosphate developing reagent" and the complex (in its reduced form) has an easily measurable blue color. The amount of phosphate present in an unknown can be estimated by comparing its color density with the color density generated by **known amounts of phosphate**. A plot of color density (in this case, colorimeter units or absorbance) as a function of known amounts of phosphate will form a "standard curve" for this comparison.

Some experimental treatments will call for the addition of the herbicide Diuron or dichlorophenyl-dimethylurea (DCMU), which is a specific inhibitor for electron transport at a site close to oxygen evolution. In other treatments, you will add NH_4Cl, an "uncoupling" reagent which inhibits phosphorylation by dissipating the pH gradient across the membrane, while allowing electron-transport to continue. Use automatic pipettors and take care when using these reagents. Figure 5-1 illustrates the expected patterns of electron flow and the sites of inhibition and uncoupling.

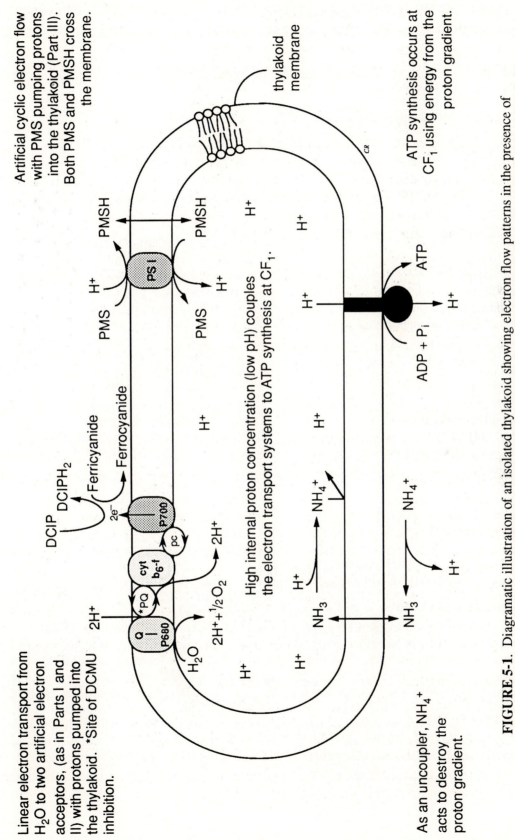

Linear electron transport from H_2O to two artificial electron acceptors, (as in Parts I and II) with protons pumped into the thylakoid. *Site of DCMU inhibition.

Artificial cyclic electron flow with PMS pumping protons into the thylakoid (Part III). Both PMS and PMSH cross the membrane.

thylakoid membrane

High internal proton concentration (low pH) couples the electron transport systems to ATP synthesis at CF_1.

ATP synthesis occurs at CF_1 using energy from the proton gradient.

As an uncoupler, NH_4^+ acts to destroy the proton gradient.

FIGURE 5-1. Diagramatic illustration of an isolated thylakoid showing electron flow patterns in the presence of three electron acceptors. The mechanism of action of the uncoupler ammonium chloride is illustrated at the lower left.

37

MATERIALS

PART I
Equipment
colorimeter with red filter
   or spectrophotometer
appropriate tubes or cuvettes
light source and heat filter
light filters (screens)
test tube racks
automatic pipettors and tips

Solutions
DCIP reaction mixture:
   0.07 mM DCIP
   0.1 M KCl
   20 mM Tricine, pH 7.5
0.01 mM DCMU NOTE: toxic, handle carefully
30 mM NH_4Cl

Cleanup: waste container for DCIP

PART II
Equipment
Clark type O_2 electrode
water jacketed reaction cell
water bath at 25° C
magnetic stirrers and fleas
light source and heat filter
automatic pipettors and tips

Solutions
ferricyanide solution: K ferricyanide (0.33 g/L),
   KCl (7.4 g/L), Tricine (3.58 g/L), pH 7.5
0.01 mM DCMU NOTE: toxic, handle carefully
30 mM NH_4Cl

PART III
Equipment
test tubes, 12 x 75 mm
test tube racks
light source and heat filter
clinical centrifuge
automatic pipettors and tips
Parafilm
colorimeter with red filter
   or spectrophotometer
appropriate tubes or cuvettes
aluminum foil

Solutions
basal reaction mixture, pH 8.3: 100 mM Tricine,
   100 mM NaCl, 10 mM $MgCl_2$, 10 mM sodium
   iso-ascorbate, 3 mM KH_2PO_4
15 mM ADP, pH 8.0
0.5 mM phenazine methosulfate (PMS)
0.01 mM DCMU NOTE: toxic, handle carefully
30 mM NH_4Cl
KH_2PO_4 solutions (0.5, 1.0, 1.5 and 2.0 mM)
10% trichloroacetic acid (TCA)
phosphate developing reagent

Cleanup: waste container for phosphate reagent

PARTS I, II and III: Plant Material
Spinach (*Spinacia oleracea*) chloroplast thylakoid membranes: 30 g of deveined spinach leaves were washed, then put in a Waring blender with 100 mL of an ice-cold solution of "SCHB," containing 20 mM HEPES buffer at pH 7.8, 400 mM sucrose as an osmoticum, 200 mM choline-Cl for ionic strength and 2 mg/mL fatty acid-free bovine serum albumin to absorb free fatty acids which might uncouple electron transport. The leaves were ground for 10 to 15 seconds at high speed. The cold homogenate was kept on ice and strained through cheesecloth to remove unbroken tissue. The homogenate was then centrifuged 5 minutes at 1900 *g*. The resulting supernatant was discarded and the pellet resuspended in SCHB and centrifuged at 1900 *g* for 5 minutes to sediment thylakoids. These were resuspended in 10 mL in fresh buffer. The chlorophyll concentration was determined by extracting an aliquot with 80% acetone, and measuring its color density at the chlorophyll peaks. The thylakoid preparation was diluted to a final concentration of 0.5 mg/mL chlorophyll and kept on ice.

PART I. The Reduction of Indophenol Dye. In Experiment A you will determine a standard reaction time for reduction of the dye, which will be used as the light exposure time for Experiments B, C, D and E.

PROCEDURE

Before starting, zero the colorimeter or spectrophotometer (set at 600 nm) with a water blank.

During the experiment, position the appropriate test tubes or cuvettes in the same place with respect to the light source for each light exposure. Make the colorimeter/spectrophotometer readings rapidly so that the light from the instrument does not significantly reduce the dye. Try to keep the tubes free of fingerprints. NOTE: Dispose of all solutions containing DCIP in the proper waste container.

A.   Time-course. You will make repeated measurements of the color density of a mixture of dye and thylakoids until all of the dye is used, then determine the light exposure time that causes reduction of one-half to two-thirds of the DCIP. This light exposure time will be the standard reaction time for Experiments B, C, D and E.

(1)   Fill one colorimeter tube or cuvette with 5 mL of the DCIP reaction mixture, add 0.1 mL of the thylakoid preparation and mix. Immediately measure the color density. Make your reading quickly so that the dye is not reduced in the instrument and go to step (2) immediately. NOTE: If small cuvettes are used, add a smaller amount of DCIP.

(2)   Expose the mixture to bright light (filtered through a water bath as a heat sink) for 15 seconds, then read the color density again. Expose the mixture to bright light for repeated 15 second intervals followed by a measurement of the color density, until the reaction is largely over (i.e. when there is no further change in the readings). Since the reaction mix contains thylakoids and other components, the final reading will **not** be zero. Record the reading taken after each time exposure in the chart on page 45.

CALCULATIONS

Plot the color density (colorimeter units or absorbance) vs. the time of illumination on the graph provided (page 45). Determine the time of illumination which causes one-half to two-thirds the loss of blue color; this will be your standard illumination time for the rest of the experiments in Part I.

B.   Light Intensity Series. Measurement of DCIP reduction at different light intensities. For each experiment (B-E), fill the required number of tubes with 5 mL of the

DCIP reaction mixture. Next add any special components such as inhibitors. Finally, put in a 0.1 mL aliquot of the thylakoid preparation (which has been kept on ice) in one tube and mix by inverting it with a piece of Parafilm over the top. Read the initial color density. Expose the tube to light for your determined reaction time, then take a second reading. The **difference** between the two readings is proportional to the amount of dye reduced. Record your data for Experiments B-E in the charts on pages 46 through 47.

(1)   Fill five colorimeter tubes or cuvettes with 5 mL of the DCIP reaction mixture.

(2)   Add 0.1 mL of the thylakoid preparation to the first tube only.

(3)   Rapidly measure the color density.

(4)   Expose the mixture to full light intensity for the standard time (as determined in Experiment A).

(5)   Read the color density again.

(6)   Add 0.1 mL of the thylakoid preparation to each remaining tube, one at a time. Measure the color density and run the reaction for the standard time at a different light intensity each time (50%, 30%, 10%, or 5% of the original light), using specially calibrated plastic wire screens. Make another reading immediately after each exposure.

C.   Light Intensity Series in the Presence of DCMU. Repeat Experiment B, but add 0.1 mL of 0.01 mM DCMU to each tube. **Use a micropipet to add this herbicide and be careful.** The final concentration of DCMU will be 2×10^{-7} M.

D.   Full Light Reaction in the Presence of an Uncoupler. Repeat the basic reaction **at full light intensity only,** but add 0.5 mL of 30 mM NH_4Cl to the tube. **Use a pipetting device.** The final concentration of NH_4Cl will be 3 mM.

E.   Will Boiled Thylakoid Membranes Reduce DCIP? Place a tube containing 0.1 mL of the chloroplast preparation in boiling water for 1 minute. Then add the DCIP reaction mix and repeat the basic reaction **at full light intensity only.**

CALCULATIONS

Determine the Δ colorimeter units or Δ absorbance for each treatment. Plot the amount of dye reduced (Δ colorimeter units or Δ absorbance) vs. light intensity for Experiments B and C on the graph provided. Put the point values from Experiments D and E on the graph too.

PART II. Measurement of the Effect of an Inhibitor and an Uncoupler on Oxygen Evolution. The oxygen electrode will be used to measure O_2 concentration. Calibrate the O_2 electrode (see Expt. 7).

PROCEDURE

(1)   Add 2.0 mL of the ferricyanide solution to the reaction cell.

(2)   When you are ready to begin, pipet 0.1 mL of the thylakoid preparation into the reaction cell and replace the cap. Once the oxygen electrode measures a steady level of oxygen in the dark (if O_2 is being consumed, correct all future rates of O_2 evolution), turn on a bright light and let the reaction proceed until a steady rate of oxygen evolution is observed. This will be your control rate of O_2 evolution.

(3)   Use a pipeting device to add 0.2 mL of 0.01 mM DCMU and determine the rate of oxygen evolution. **Use a pipetting device and be careful**; although you are using only a very small quantity, you should be aware that DCMU is a toxic herbicide.

(4)   Use a fresh reaction mixture (repeat steps 1 and 2), then add 0.2 mL of 30 mM NH_4Cl and determine the rate of oxygen evolution in the presence of this uncoupler. **Be sure to use a pipetting device.**

CALCULATIONS

Determine the rates in units/min and record the data in the chart on page 47. Follow the directions in the Calculations section of Expt. 7 and determine the rate of O_2 evolution in $\mu mol/mL$-h for each treatment. Considering that the final concentration of chlorophyll in the reaction cell was 0.025 mg/mL, determine the rate of O_2 evolution in $\mu mol/mg$-h.

PART III. **Photosynthetic Phosphorylation**. Measurement of the amount of inorganic phosphate which remains at the end of the reaction.

PROCEDURE

(1)   Prepare 12 small test tubes (12 x 75 mm) for the reaction mixtures (treatments 1-6), the background reading (7), and the standard curve (treatments 8-12) according to the protocols in Tables 5-1 and 5-2. Wrap the test tube for treatment 2 in aluminum foil. Do not add the thylakoids until you are ready to begin.

The first six test tubes are for the experimental treatments. The "basal" mixture for these treatments contains several components at pH 8.3 (see Materials).

TABLE 5-1. Protocol for Treatments 1- 6, Part III.

Tube #:	1	2	3	4	5	6
Component	Control	Dark*	– ADP	+ DCMU	+ NH$_4$Cl	Minus Thylakoids
Basal Mixture	0.5 mL	0.5 mL	0.5 mL	0.5 mL	0.5 mL	0.5 mL
15 mM ADP	0.1	0.1	0	0.1	0.1	0.1
0.5 mM PMS	0.1	0.1	0.1	0.1	0.1	0.1
0.01 mM DCMU**	0	0	0	0.1	0	0
30 mM NH$_4$Cl	0	0	0	0	0.1	0
H$_2$O	0.2	0.2	0.3	0.1	0.1	0.3
Thylakoids***	0.1	0.1	0.1	0.1	0.1	0

* Tube 2 will be the "dark" control. It should be wrapped in aluminum foil before the thylakoids are added.

** Final concentration of DCMU will be 10^{-6} M.

*** DO NOT add the thylakoids until you are ready to start the experiment.

Test tube 7 will be used as the background measurement and will account for any phosphate that may be added to the mixture from the thylakoid preparation. The final reading for this treatment will be subtracted from the readings for tubes 1 to 5.

Tubes 8 to 12 will comprise a standard curve for phosphate concentration. The solutions for the standard curve have been mixed in half strength basal solution (without the 3 mM phosphate) to provide the same final concentrations of all **other** ingredients found in your reaction mixtures; phosphate was then added to give the concentrations listed in Table 5-2. The solution used for tube 8 contains half strength basal solution, but contains **no** phosphate. The concentration of phosphate present in tube 11 (1.5 mM KH$_2$PO$_4$) is the same as the starting concentration of inorganic phosphate in your reaction tubes (1-6).

(2) Add the thylakoids to tubes 1 to 5 (but **not** 6). Immediately place tubes 1, 3 to 6 and tube 7 in a test tube rack at full light intensity for 5 minutes (the reaction time).

TABLE 5-2. Protocol for Tubes 7-12, Part III.

#	Component	Volume
7	Thylakoid Preparation	0.1 mL
	+ Distilled H$_2$O	0.9 mL
8	0 mM KH$_2$PO$_4$	1.0 mL
9	0.5 mM KH$_2$PO$_4$	1.0 mL
10	1.0 mM KH$_2$PO$_4$	1.0 mL
11	1.5 mM KH$_2$PO$_4$	1.0 mL
12	2.0 mM KH$_2$PO$_4$	1.0 mL

(3)   Turn off the light and add 0.2 mL of 10% trichloroacetic acid (TCA) to **all** of the test tubes (including 2 and 7-12) to stop the reaction. **Be sure to use a pipetting device; TCA can burn your skin.**

(4)   After two minutes, centrifuge **all** the tubes for five minutes in a clinical centrifuge at sub-maximal speed to sediment flocculated, denatured thylakoid membranes.

(5)   Put 4.0 mL of the phosphate developing reagent into each of twelve colorimeter or spectrophotometer tubes. Then transfer 0.9 mL of supernatant (using a new pipet tip for each) into each appropriately labeled tube. Invert to stir or use a vortex mixer. NOTE: If a small cuvette is needed for your spectrophotometer, transfer a measured portion of each of the final mixtures to the cuvette to take each reading.

(6)   Wait at least one minute before measuring the color density in the colorimeter (red filter) or spectrophotometer (600 nm), which has been calibrated with a water blank. Dilute all the treatments by a measured amount if the solutions are too dense to measure easily. Record the colorimeter units or absorbance for each tube in the chart on page 48.

CALCULATIONS

Construct a "standard curve" by plotting your readings for tubes 8 to 12 against the **amount** of phosphate in these tubes on the graph provided (page 49). Be sure to convert concentration to amount.

Subtract the value for tube 7 from the values for tubes 1 to 5. Use the chart on page 48. Determine the approximate μmoles of inorganic phosphate left in each reaction tube after illumination by comparing your corrected readings for tubes 1 to 6 to the standard curve. Subtract the amount left from the total originally present (1.5 μmol) to find the amount of phosphate esterified into ATP.

TEXT REFERENCES

Galston, A. W., P. J. Davies and R. L. Satter, *The Life of the Green Plant* (3rd ed.), pp. 92-97. Englewood Cliffs, NJ: Prentice Hall, 1980.

Salisbury, F. B. and C. W. Ross, *Plant Physiology* (4th ed.), pp. 214-224. Belmont. CA: Wadsworth Publishing Co., 1991.

Taiz, L. and E. Zeiger, *Plant Physiology*, pp. 204-216. Redwood City, CA: Benjamin/Cummings Publishing Co., Inc., 1991.

FURTHER READING

Arnon, D. I., "The discovery of photosynthetic phosphorylation," *Trends in Biochemical Sciences* (1984), 9:258-262.

Clayton, R. K., *Light and Living Matter: A Guide to the Study of Photobiology*, Vol 2. New York: McGraw-Hill, 1971.

Dennis, D. T., *The Biochemistry of Energy Utilization in Plants*. New York: Chapman and Hall, 1987.

Govindjee and W. J. Coleman, "How plants make oxygen," *Scientific American* (1990), 262(2):50-58.

Gregory, R. P. F., *Photosynthesis*. New York: Chapman and Hall, 1989.

Haehnel, W., "Photosynthetic e⁻ transport in higher plants," *Annu. Rev. Plant Physiol.* (1984), 35:659-693.

Hinkle, P. C. and R. E. McCarty, "How cells make ATP," *Scientific American* (1978), 238(3):104-123.

EXPERIMENT 5

Name _____

Date _____

RESULTS

PART I, A.

Record your results in the table below.

Time										
Colorimeter Units or A_{600}										

Plot colorimeter units or absorbance vs. time for Part A on the graph below, then determine the standard reaction time.

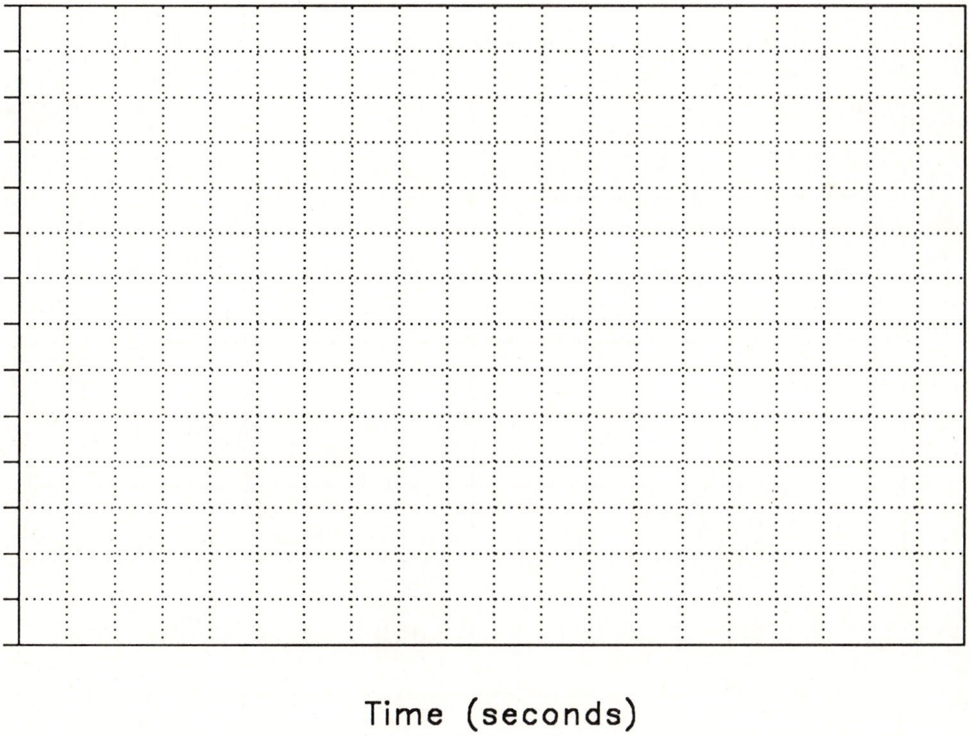

Time (seconds)

Standard Reaction Time _____

PART I, B-E.

Record your results (as colorimeter units or absorbance) in the tables that follow.

B.   Control light intensity series.

% Light	Initial Units or A_{600}	Final Units or A_{600}	Δ Units or ΔA_{600}
100			
50			
30			
10			
5			

C.   Light intensity series in the presence of DCMU.

% Light	Initial Units or A_{600}	Final Units or A_{600}	Δ Units or ΔA_{600}
100			
50			
30			
10			
5			

D.   Reduction of dye in the presence of NH_4Cl at full light intensity.

% Light	Initial Units or A_{600}	Final Units or A_{600}	Δ Units or ΔA_{600}
100			

E.   Reduction of dye using boiled thylakoids at full light intensity.

% Light	Initial Units or A_{600}	Final Units or A_{600}	Δ Units or ΔA_{600}
100			

Plot the amount of dye reduced (Δcolorimeter units or ΔA_{600}) vs. light intensity for Parts B-E.

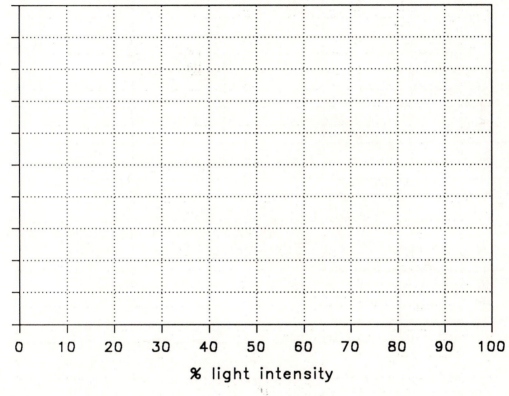

% light intensity

PART II.

Record the rates as units/min in the chart below, then determine the rates of O_2 evolution in μmol/mg-h.

Treatment	Rate (units/min)	x O_2/unit = μmol/mL/min	x 60 min/h = μmol/mL-h	O_2 Evolution (μmol/mg-h)
Control				
+ DCMU				
+ NH$_4$Cl				

PART III.

Record your raw data in the charts below.

#	Treatment	Colorimeter Units or A_{600}	– # 7 = Net Units or A_{600}	Initial Pi (µmol)	Final Pi (µmol)	µmol Pi Used (ATP made)
1	Light			1.5		
2	Dark			1.5		
3	– ADP			1.5		
4	+ DCMU			1.5		
5	+ NH₄Cl			1.5		
6	– Thylakoids			1.5		

7	Thylakoids Only	

Subtract the value for tube 7 from the values for tubes 1 to 5.

8	0 mM Pi	
9	0.5 mM Pi	
10	1.0 mM Pi	
11	1.5 mM Pi	
12	2.0 mM Pi	

Construct a standard curve by plotting your readings for tubes 8 to 12 against the **amount** of phosphate in these tubes (use the graph on the next page).

Convert your corrected colorimeter readings to µmol of phosphate (final Pi) by finding the points for tubes 1 to 6 on the graph. Then complete the chart above by calculating the µmol of ATP produced.

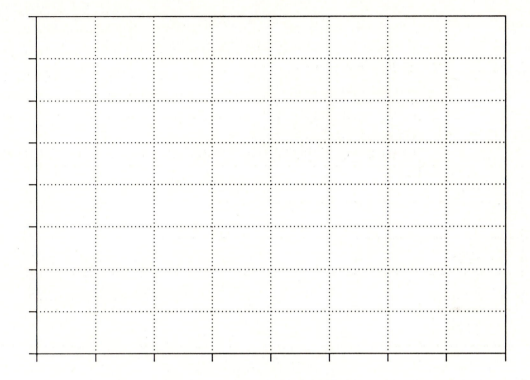

Amount of phosphate/tube (μmol)

QUESTIONS

1. Why are artificial electron acceptors supplied in these experiments?

2. You should have observed light-limited and light-saturated portions of the control light curve in Part I. Describe the effect of DCMU on the rate of electron transport.

3. Explain the effect of NH_4Cl on rate of dye reduction at full light intensity (Part I).

4. How do you know that the dye DCIP is not the limiting factor at full light intensity (light saturation) in the control curve of Part I? What is the limiting factor at full light intensity in the control?

5. Why are your results for Part I and II similar? (Don't comment on the inhibitors.)

6. From your data in Part III, what **substrates** are needed for photophosphorylation in isolated thylakoid membranes?

7. Why doesn't DCMU inhibit phosphorylation in this Part III of this experiment?

8. Explain the effect of NH_4Cl on photophosphorylation (Part III).

EXPERIMENT 6

Carbon Fixation

INTRODUCTION

The Calvin-Benson cycle encompasses the steps of photosynthesis in which energy from the light reactions is used to fix CO_2 into carbohydrates in the enzyme-rich stroma of the chloroplasts. Carbon dioxide labeled with radioactive carbon ($^{14}CO_2$) was used in Calvin and Benson's experiments with the alga *Chlorella* to determine the identity of the first product of carbon fixation and to elucidate the entire pathway (Fig. 6-1). Spinach chloroplast stroma preparations will fix $^{14}CO_2$ (or even $H^{14}CO_3^-$) into phosphoglyceric acid when provided with the essential substrates and a source of energy (ATP). In this experiment, you will supply the necessary ingredients for carbon fixation in the presence of radioactive CO_2, then monitor the products of the reaction for radioactivity. If all of the required substances are present, there should be a large amount of detectable radioactivity in the product solution. If a required substance is missing, however, not much radioactive carbon will be incorporated. In particular, you will test the requirements for a suitable 5-carbon sugar, ATP, and Mg^{+2}. The relative amounts of $^{14}CO_2$ fixed into carbohydrate materials will be determined by liquid scintillation counting.

The Liquid Scintillation Spectrometer takes advantage of the conversion of radioactive particle energy into light rays (photons) by special fluorescent chemicals, or "fluors." The radioactive compound and the fluor are mixed in an intimate way in a special, usually nonaqueous, solution. When a radioactive particle (in this case, a beta particle) is emitted, its chance of being absorbed by a molecule of the fluor is extremely good because of the intimate mixing. When some of the radioactive energy is absorbed, the fluor goes into an energetic "excited state"; the excited state decays to give off a quantum of visible light. The visible light is absorbed by a highly sensitive photomultiplier, a light-sensitive surface which triggers an amplified electric impulse when hit by light. The electric impulses are counted over a set period of time and data is generated as counts per minute (CPM). The radioactive samples for the class may be left to count overnight and the results posted on the following day.

MATERIALS

Equipment
latex gloves
lab coats
"radioactive" tape
5-mL Warburg flasks
filter paper
rubber serum stoppers to fit flasks
plastic backed absorbent paper
plastic trays
automatic pipettors
disposable pipet tips
petroleum jelly and applicator sticks
N_2 source (with attached tubing
   and syringe needle*)
1-mL plastic syringes*
18-G needles*
disposable transfer pipets
test tube holders
disposable test tubes, 12 x 75 mm
clinical centrifuge
nitrile or toluene resistant gloves
5-mL Multi scintillation vials in tray packs
Repipet containing scintillation cocktail
Liquid Scintillation Spectrometer
refrigerated centrifuge for stromal preparation

Solutions
Prepared reaction mixes:
Complete: 15 mM $MgCl_2$, 5 mM Ri-5-P, 10
   mM ATP in suspension medium, pH 8
– ATP: 15 mM $MgCl_2$, 5 mM Ri-5-P
   in suspension medium
– Ri-5-P: 15 mM $MgCl_2$, 10 mM ATP
   in suspension medium
– Mg^{2+}: 5 mM Ri-5-P, 10 mM ATP
   in suspension medium
0.2 M ATP, pH 7 (add 1 mL to mixes 1, 3 and 4
   at the beginning of the experiment)
10% perchloric acid
30% NaOH
Scintillation cocktail (see Appendix G)
95% ethyl alcohol
$NaH^{14}CO_3$ (3.125 μCi/mL) in 0.02 M $NaHCO_3$
   stored in secondary containment

Cleanup
Q-tips
liquid radioactive waste container
solid radioactive waste container
plastic container of 30% NaOH
   for wick disposal

Plant material
Spinach (*Spinacia oleracea*) chloroplast preparation: Sixty grams of deveined spinach were ground in a blender with 100 mL of cold homogenization medium containing 50 mM Tris (pH 7.8) as a buffer, 0.01 KCl for ionic strength and 400 mM sucrose as an osmoticum, then filtered through a double layer of cheesecloth. The filtrate was then sedimented at 120 *g* for 8 minutes to remove cell debris. The supernatant was again sedimented, but at 1075 *g* for 8 minutes to pellet whole chloroplasts. The resulting supernatant was discarded and the pellet was resuspended in 8 mL of buffered suspension medium without sucrose, to pop open the chloroplast envelope. The preparation must be kept on ice.

*NOTE: The distribution of syringes and hypodermic needles is controlled by law in most states. Be sure to follow your instructor's advice about the storage and disposal of syringes.

IMPORTANT: Great care should be taken when using radioactive isotopes to avoid transferring any isotopes to your hands, clothes, or equipment. The following precautions are mandatory when a low-energy β radiation emitter like ^{14}C is used. PLEASE OBSERVE THESE PRECAUTIONS FOR THE HANDLING OF RADIOACTIVE MATERIALS DURING CLASS!

1. Never pipet radioactive solutions by mouth. Pipetting devices must be used to transfer radioactive solutions.
2. Carry out all work in a tray lined with absorbent paper so that any minor spills will be contained. Keep stock solutions of ^{14}C in secondary containment as well.
3. Wear latex gloves and a disposable lab coat when radioactive solutions are used.
4. Label all materials which come in contact with the radioactive solution with warning tape. Remove the tape when these items are decontaminated.
5. Dispose of waste radioactive materials in the special containers provided, but do not put other, non-radioactive, materials in these containers.
6. Tell your instructor of any accidental spillage of radioactive solution or materials!
7. Read Appendix E about radioactivity.
8. The instructor will conduct wipe tests (see Appendix E) at the end of the class to ensure that no area of the laboratory is contaminated.

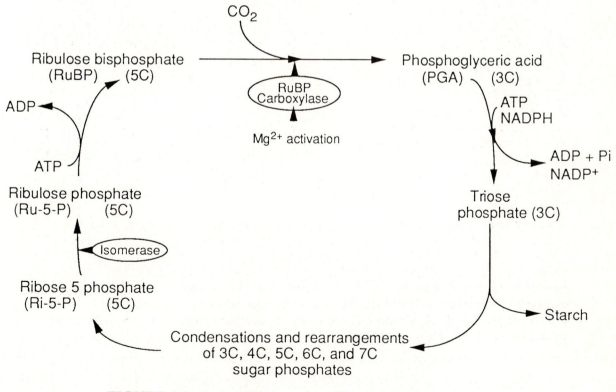

FIGURE 6-1. A simplified version of the Calvin-Benson Cycle.

PROCEDURE

The components needed for carbon fixation will be placed in a Warburg flask with ^{14}C-labeled bicarbonate. The chloroplast mixture is added to start the reaction, which is stopped at the end of 10 minutes with the addition of an acid solution. Lowering the pH causes the unfixed $^{14}CO_2$ to be driven off; a sample of the remaining mix may then be tested for radioactivity.

(1)   Obtain four Warburg flasks and make sure they are clean. Use a Q-tip to remove any visible residue.

(2)   Label the flasks and (after the instructor has added the ATP to three of the mixtures) place 2 mL of the of the appropriate reaction mixture into the main chamber of each flask (Fig. 6-2).

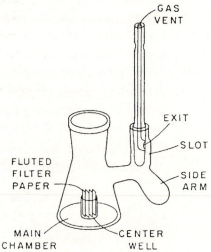

FIGURE 6-2. The Warburg flask.

     1.   Complete treatment: contains suspension medium, 10 mM ATP, 5 mM ribose-5-phosphate (Ri-5-P is converted by isomerase in the chloroplast extract to ribulose-5-phosphate or Ru-5-P) and 15 mM $MgCl_2$
     2.   – ATP
     3.   – Ri-5-P
     4.   – Mg^{2+}

(3)   Grease the upper edge of the center well in each flask with petroleum jelly to prevent mixing of the contents of the well with the reaction mix in the main chamber. Cut four squares (about 4 x 2 cm) of filter paper and fold each into accordion pleats. With forceps, place one fluted wick of filter paper into the center well of each flask.

(4)   Add 0.3 mL of the chloroplast suspension to the side arm of each flask.

(5)   Ask the instructor to add 0.03 mL of $NaH^{14}CO_3$ to the main chamber of each flask. Cap the flask with a greased rubber serum stopper.

(6)   At the fume hood, flush each flask with N_2 for 30 seconds (push the syringe needle through the serum cap) with the side arm stopper removed, then replace the side arm stopper, making sure that the vent is not in the open position. Gassing with N_2 removes the oxygen, which could decrease the effectiveness of RUBP carboxylase in fixing carbon.

(7)   Note the time, then turn the flask on its side and **roll** the chloroplasts into the main chamber. Make sure the contents have mixed completely, then let the reaction run for 10 minutes.

(8)   Just before the reaction time is up, use a syringe to puncture the serum cap and add 0.2 mL of 30% NaOH to the **center well**. Use care and report any needle stick injury to your instructor. The NaOH will absorb any $^{14}CO_2$ which might be released during the next step (9).

(9)   At the end of the 10-minute reaction time, again puncture the serum cap with a syringe to add 0.3 mL of 10% perchloric acid to the **main chamber**. Report any spills of perchloric acid to your instructor. The acid will stop the reaction by precipitating proteins and will drive off any unfixed CO_2. **Leave the capped flasks in your tray for 30 minutes**, gently shaking the tray periodically. All of the **unfixed** $H^{14}CO_3$ will be converted to $^{14}CO_2$, which will be released as a gas from the reaction mix and absorbed by the NaOH in the center well.

In the main chamber: $H^{14}CO_3^- + H^+ \rightarrow {}^{14}CO_2 + H_2O$ Eq. (6-1)

In the center well: $NaOH + {}^{14}CO_2 \rightarrow NaH^{14}CO_3$ Eq. (6-2)

(10)   After 30 minutes, move to the fume hood and remove the serum caps. Using a new transfer pipet each time, transfer the solution in each main chamber to a labeled centrifuge tube (use disposable plastic 12 x 75 mm tubes). Keep the used pipets in your tray until you are able to dispose of them properly.

(11)   Cap the test tubes and sediment the solutions in a clinical centrifuge for two minutes to remove the chloroplast debris.

(12)   Obtain 10 plastic scintillation vials and label the caps. Two vials will be used for each treatment and two will be used to determine background counts. Wearing toluene resistant gloves, add 3 mL of scintillation cocktail to each vial. Be careful. Scintillation fluid should be treated with as much care as radioactive materials.

(13)   With an automatic pipetting device and a new plastic tip for each treatment, take two 0.25 mL-samples of supernatant from each centrifuge tube and place each sample into the appropriate vial. Add 0.25 mL of H_2O to each of the background vials. The background vials will give readings of counts per minutes attributable to other sources of radiation which might be present; these readings will be subtracted from all the other readings. Cap the vials firmly and shake each vial to mix the contents.

The instructor will collect the vials and begin the scintillation counting. The radioactive counts for your samples will be measured in the Liquid Scintillation Spectrometer. The spectrometer will automatically record the counts for each vial; the instructor will inform you where and when the results will be available.

CLEANUP

All radioactive materials should be placed in either the liquid or solid radioactive waste according to the lists below. Once the flasks have been rinsed twice with 95% ethyl alcohol, they may be washed as usual. Use Q-tips and ethyl alcohol to remove any remaining residue.

Solid waste	Liquid waste
pipet tips	solutions from the
Pasteur pipets	reaction vessels
centrifuge tubes	first and second rinses
Q-tips used for cleanup	of the reaction vessels
wicks (in the container provided)	

CALCULATIONS

The counts per minute (CPM) should be averaged for each treatment and then corrected for background. Use the chart in the Results section.

TEXT REFERENCES

Galston, A. W., P. J. Davies, and R. L. Satter, *The Life of the Green Plant* (3rd ed.), pp. 97-100. Englewood Cliffs, NJ: Prentice Hall, 1980.

Salisbury, F. B. and C. W. Ross, *Plant Physiology* (4th ed.), Chap. 11. Belmont, CA: Wadsworth Publishing Co., 1991.

Taiz, L. and E. Zeiger, *Plant Physiology*, pp. 219-228. Redwood City, CA: Benjamin/Cummings Publishing Co., Inc., 1991.

FURTHER READING

Bassham, J. A. and M. Calvin, *The Path of Carbon in Photosynthesis*. Englewood Cliffs, NJ: Prentice Hall, 1957.

Calvin, M., "Forty years of photosynthesis and related activities," *Photosynthesis Research* (1989), 21:1-16.

Ellis, T. J., "The most abundant protein in the world," *Trends in Biochemical Sciences* (1979), 4:241-244.

Jensen, R. G. and J. T. Bahr, "Ribulose bisphosphate carboxylase/oxygenase," *Annu. Rev. Plant Physiol.* (1977), 28:379-400.

Woodrow, I. E. and J. A. Berry, "Enzymatic regulation of photosynthetic CO_2 fixation in C_3 plants," *Annu. Rev. Plant Physiol. Plant Mol. Biol.* (1988), 39:533-594.

EXPERIMENT 6

Name _____

Date _____

RESULTS

Record the CPM in the chart below and calculate the corrected CPM for each treatment.

#	Treatment	CPM		Average CPM	– Background = Corrected CPM
1	Complete				
2	– ATP				
3	– Ri-5-P				
4	– Mg^{2+}				
	Background				

QUESTIONS

1.   How do your results for treatments 2, 3 and 4 compare with the control? Discuss what each treatment tells you about the requirements for carbon fixation.

2.   Where does the Ri-5-P come from *in vivo*?

3.   Why isn't light required for this experiment? What role does the chloroplast preparation play in this experiment?

4.   What additional control(s) might be added to this experiment to help answer question 3?

5.   Why did you flush the flasks with N_2? Explain what effect the presence of oxygen has on carbon fixation.

EXPERIMENT 7

Light Relations in Whole Cell Photosynthesis

INTRODUCTION

Light should be considered as one of the substrates for photosynthesis, since it is used in an amount stoichiometric to the other substrates and to the generation of products. The relationship between rate and light intensity, therefore, is one of the most fundamental characteristics of any photosynthetic system. If the rate of O_2 evolution is measured as a function of light intensity, two regions in the light curve should be apparent: the region where light is limiting and the region of light saturation where further increases in the abundance of light energy or quanta have no further effect on the rate. At light saturation, some other factor is limiting the rate of photosynthesis, most likely one of the enzymatic steps of the carbon reduction or Calvin-Benson cycle (Fig. 7-1).

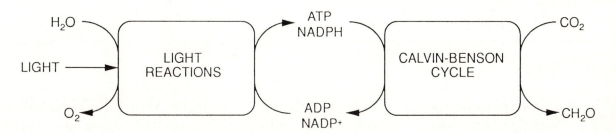

FIGURE 7-1. The interaction between the light reactions and the Calvin-Benson cycle.

Inhibitory agents that cause quanta to be wasted will act most strongly on the light-limited part of the curve, where an increase in rate is possible only with an increase in effectively used light. Treatments that inhibit the enzymatic reactions (the Calvin-Benson cycle) of photosynthesis will have a greater effect when the rate of O_2 evolution is light saturated. The herbicide Diuron or dichlorophenyl-dimethylurea (DCMU) is a specific inhibitor for electron transport at a site close to oxygen evolution. The rate of O_2 evolution will be measured in the presence of DCMU and at low temperature; the effect of each on the light intensity curve will be determined, providing information about the nature of the interaction between the light reactions and the Calvin-Benson cycle.

The rate of O_2 evolution may be determined with an O_2 electrode, which measures the concentration of dissolved O_2 in the reaction cell (see Appendix D). The unicellular alga *Chlorella vulgaris* var. viridis will be the organism used for the study of photosynthesis in whole cells.

MATERIALS

Equipment
O_2 electrode with teflon membrane
water jacketed reaction cell
water bath at 25°C
magnetic stirrer
lamps and heat filter
light filters (screens)
aspirators fitted with flexible plastic tips
Pasteur pipets and bulbs
chart recorders
automatic pipettors
12 x 75 mm centrifuge tubes
clinical centrifuge
aluminum foil
small artist's brush

Solutions
sodium dithionite (sodium hydrosulfite)
Distilled H_2O, 25°C and 15°C
95% ethyl alcohol
20 μM DCMU
0.1 M K_2CO_3-$NaHCO_3$ buffer

Plant material
Chlorella vulgaris var. viridis
preparation in 0.1 M K_2CO_3-$NaHCO_3$
buffer at pH 8.9.

PROCEDURE

A standard suspension of *Chlorella* cells will be provided in a K_2CO_3-$NaHCO_3$ buffer. Photosynthesis will be measured by monitoring O_2 evolution using the oxygen electrode system, operating at 25°C.

(1)   The recorder must first be calibrated. Turn on the polarizing voltage, magnetic stirrer and the recorder. Make sure the water bath is at 25°C. (NOTE: the following instructions are for the Clark electrode equipped as described in Appendix H. If you have a more advanced model, the use of dithionite may not be necessary. Your instructor will advise you on how to proceed.) Add a small pinch of sodium dithionite to the distilled H_2O in the reaction cell, cap the cell, and adjust the zero control knob so that the recorder pen lies between 0 and 10 on the chart. The sodium dithionite will remove all dissolved O_2 from the water:

$$Na_2S_2O_4 + O_2 + H_2O \rightarrow NaHSO_4 + NaHSO_3 \quad \text{Eq. (7-1)}$$

Now aspirate out the dithionite solution, rinse three times, and replace with fresh distilled water at 25 °C (keep a distilled H_2O bottle in the water bath and shake it vigorously before adding fresh water to the reaction cell). The water in the reaction cell must be in equilibrium with air, so leave the cell uncapped for this part of the calibration. Adjust the sensitivity control knob so that the pen lies between 90 and 100 on the chart. Note that this knob changes the degree of pen deflection per the concentration of oxygen in the water.

Now add dithionite again and recheck the zero point. It may be necessary to re-adjust the zero knob. Go back and forth between water and dithionite as necessary, until the pen travels a consistent number of chart units for the difference between air-saturated water and water with dithionite. If the pen goes off-scale on the high side, as it may when water is in the reaction cell, bring it back on-scale with the sensitivity knob. If off-scale on the low side, bring it back on with the zero control knob. When the calibration is completed, note the number of chart units between the two extremes. Once the electrode is calibrated, the zero control knob may be adjusted at any time during the experiment, but the sensitivity setting **must not** be changed. Before beginning make sure that the light is in the correct position to be focused on the reaction cell, then turn the lamp off.

(2)   Place a standard suspension of *Chlorella* in the reaction cell and replace the cap (rinse the cap with distilled water first). Cover the reaction cell with foil and measure the rate of O_2 evolution in the dark. Let the reaction run until a straight line with a measurable slope is seen. Record the slope and remove the foil.

(3)   Place a neutral screen which transmits about 5% of the incident light between the light source and the reaction cell, then switch on the lamp. Mark the starting point on the chart. Continue illumination until you see a straight line, with an easily measurable slope (depending on internal concentrations of intermediates, photosynthetic cells may show a series of two or three differing transients or fast variations in rate for the first minute or so after the light goes on).

(4)   Repeat step (3), but use filters that transmit 10, 30 and 50% of the incident light. Finally, determine the rate of oxygen evolution at full light intensity. These six treatments will provide the data for your light intensity curve.

(5)   Fill the reaction cell half full with fresh cell suspension, then add 25 μL of the electron transport inhibitor DCMU. Use an automatic pipettor and be careful when pipetting this herbicide. Fill the reaction cell to the top with additional cell suspension and replace the cap. Measure the rate of oxygen evolution in the dark

and at all light intensities once more. After using the inhibitor, rinse the reaction cell carefully with 95% ethyl alcohol and then with distilled H_2O.

(6)   Again measure the rate at all light intensities and in the dark but use a cell suspension three times as thick (sediment 6 mL of the standard suspension in a clinical centrifuge for 3 minutes, then, using a small artist's brush, resuspend the pellet in 2 mL of 0.1 M K_2CO_3-$NaHCO_3$ buffer).

(7)   Finally, repeat the procedure for the standard *Chlorella* reaction at all light intensities and in the dark at 15 °C (change the temperature of the water bath). You must recalibrate the electrode at the new temperature; be sure to use 15 °C distilled H_2O for the calibration.

CALCULATIONS

With a ruler, extend all straight lines and measure the rates. Record the rates in the charts on pages 65 through 68. Subtract the dark rate for each treatment from the rates at all light intensities. Determine the rates as μmoles of oxygen evolved per mL of suspension per hour. Follow the example below to convert corrected units/min to nmol O_2/mL-h.

Sample calculations:

1.   Pen position plus dithionite = 5 units

2.   Pen position at air saturation = 95 units

3.   Range = 90 units

4.   Temperature = 25 °C

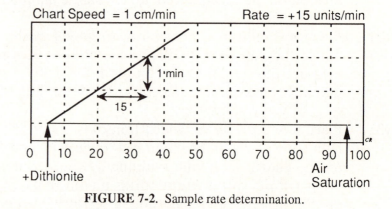

FIGURE 7-2. Sample rate determination.

5.   The concentration of O_2 in water which is in equilibrium with the air at 25 °C is 0.25 mM O_2 (see Appendix D), which is equal to 0.25 mmol/L or 0.25 μmol/mL.

6.   Therefore, **for the range defined in this example calculation**, 90 chart units = 0.25 μmol/mL OR 1 chart unit = 0.0028 μmol/mL

7.   The rate at each light intensity must be corrected for the rate in the dark. [NOTE: You may assume for this experiment that the rate of dark respiration is equivalent to the rate in the light (although the rate of **mitochondrial** respiration may change

slightly). *Chlorella* cells have a CO_2-concentrating mechanism and undergo almost no photorespiration.] If, in the dark, the pen moved to the left 1 chart unit in one minute (giving a rate of -1 unit/min), then this rate must be subtracted from the rate at each light intensity. For example, if the pen moved to the right 15 chart units in one minute in the light (as in Fig. 7-2), the corrected rate of O_2 evolution in units/min would be:

$$+15 \text{ units/min} - (-1 \text{ unit/min}) = +16 \text{ units/min}$$

8.   Therefore, **for this example**, the **corrected** rate of O_2 evolution in μmol/mL-min would be:

$$0.0028 \text{ } \mu\text{mol/mL-unit} \times 16 \text{ units/min} = 0.045 \text{ } \mu\text{mol/mL-min}$$

and for one hour:

$$0.045 \text{ } \mu\text{mol/mL-min} \times 60 \text{ min/h} = 2.7 \text{ } \mu\text{mol/mL-h}$$

% Light	Rate (units/min)	– Dark Rate = Corrected Rate (units/min)	x O_2 /unit = μmol/mL/min	x 60 min/h = μmol/mL/h
0	-1	0	0	0
100	$+15$	$+16$	0.045	2.7

9.   Determine the corrected rate of O_2 evolution **at each** light intensity, then plot the rate against light intensity on the graphs provided on pages 66 through 68.

TEXT REFERENCES

Galston, A. W., P. J. Davies, and R. L. Satter, *The Life of the Green Plant* (3rd ed.), pp. 84-44 and 100-110. Englewood Cliffs, NJ: Prentice Hall, 1980.

Salisbury, F. B. and C. W. Ross, *Plant Physiology* (4th ed.), pp. 218-226. Belmont, CA: Wadsworth Publishing Co., 1991.

Taiz, L. and E. Zeiger, *Plant Physiology*, pp. 254-263. Redwood City, CA: Benjamin/Cummings Publishing Co., Inc., 1991.

FURTHER READING

Amesz, J., "Photosynthesis," *New Comprehensive Biochemistry*, Vol. 15, Amsterdam: Elsevier, 1987.

Dennis, D. T., *The Biochemistry of Energy Utilization in Plants*. New York: Chapman and Hall, 1987.

Gregory, R. P. F., *Photosynthesis*. New York: Chapman and Hall, 1989.

Gregory, R. P. F., *Biochemistry of Photosynthesis* (3rd ed.). London: John Wiley & Sons, Ltd., 1989.

EXPERIMENT 7   Name _____

                                                Date _____

RESULTS

Record all results in the tables below. Subtract the O_2 uptake in the dark (Note: subtracting a negative number means addition), then determine the rate as μmol/mL-h.

Treatment:_____

Pen position
plus dithionite:

_____

Pen position at
air saturation:

_____

Range:

_____

Temperature:

_____

Chart speed:

_____

% Light	Rate (units/min)	− Dark Rate = Corrected Rate (units/min)	x O_2/unit = μmol/mL/min	x 60 min/h = μmol/mL/h
0				
5				
10				
30				
50				
100				

1 chart unit = _____ μmol/mL

Plot your control light intensity curve as rate of O_2 evolution in μmol/mL-h vs. light intensity (as %) on the graph on page 66. Plot the data for an additional treatment on the same graph. Be sure to label the plots for each treatment.

Treatment:_____   Chart speed: _____

Pen position
plus dithionite:

_____

Pen position at
air saturation:

_____

Range:

_____

Temperature:

_____

% Light	Rate (units/min)	– Dark Rate = Corrected Rate (units/min)	x O$_2$ /unit = μmol/mL/min	x 60 min/h = μmol/mL/h
0				
5				
10				
30				
50				
100				

1 chart unit = _____ μmol/mL

Plot your light intensity curves for the control and one other treatment on the graph below. Be sure to label the plots for each treatment.

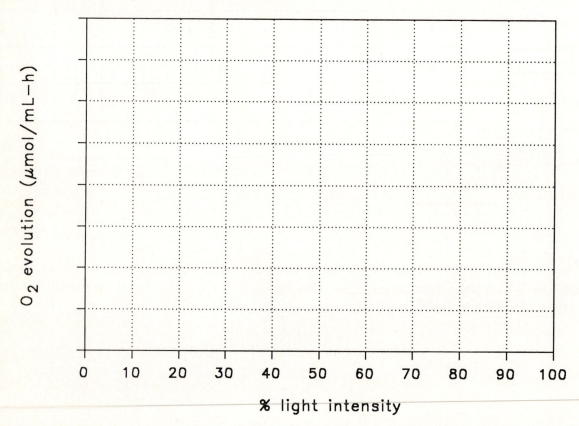

Treatment:_____   Chart speed: _____

Pen position
plus dithionite:

_____

Pen position at
air saturation:

_____

Range:

_____

Temperature:

_____

% Light	Rate (units/min)	– Dark Rate = Corrected Rate (units/min)	x O$_2$ /unit = μmol/mL/min	x 60 min/h = μmol/mL/h
0				
5				
10				
30				
50				
100				

1 chart unit = _____ μmol/mL

Plot your light intensity curves for the control and one other treatment on the graph below. Be sure to label the plots for each treatment.

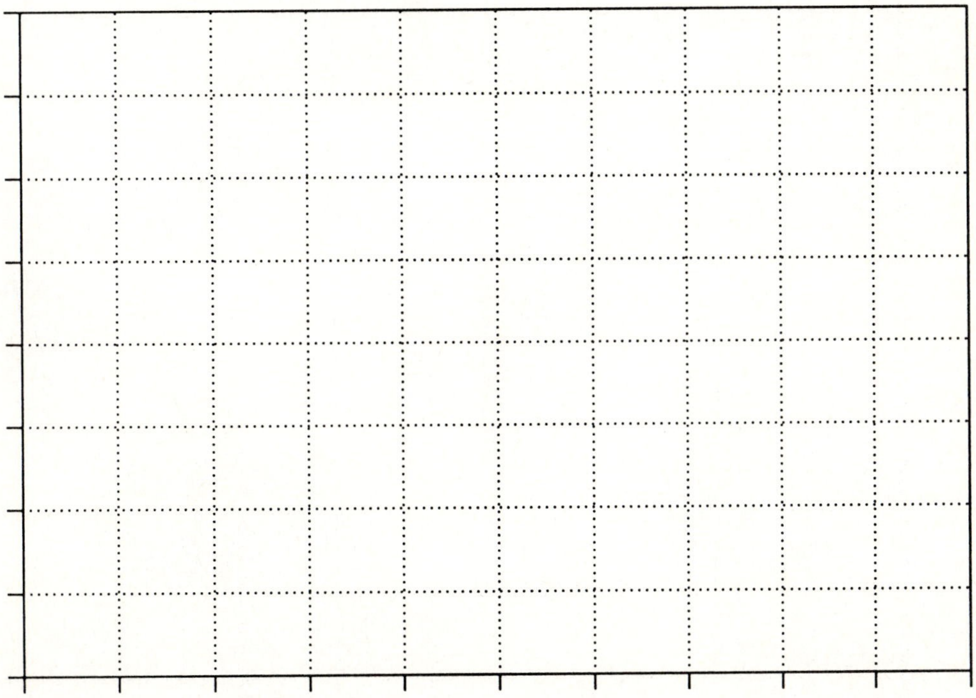

Treatment:_____   Chart speed: _____

Pen position
plus dithionite:

_____

Pen position at
air saturation:

_____

Range:

_____

Temperature:

_____

% Light	Rate (units/min)	– Dark Rate = Corrected Rate (units/min)	x O$_2$ /unit = μmol/mL/min	x 60 min/h = μmol/mL/h
0				
5				
10				
30				
50				
100				

1 chart unit = _____ μmol/mL

Plot your light intensity curves for the control and one other treatment on the graph below. Be sure to label the plots for each treatment.

QUESTIONS

1.   Why did you subtract the O_2 uptake in the dark?

2.   Describe briefly how light intensity affects O_2 evolution. Refer specifically to the graph of your control treatment. What is the limiting factor at low light intensities? At high light intensities?

3.   What effect does DCMU have on the light intensity curve and why? Account for the specific site of DCMU inhibition in your answer.

4.   How and why is the light intensity curve affected by an inhibitor of Calvin-Benson cycle enzymatic activity such as low temperature?

5.   Why is the light limiting area of the curve not affected by low temperature?

6.   How does increasing the concentration of *Chlorella* cells affect the rate of O_2 evolution? Does the rate of O_2 evolution vary proportionately with the concentration of *Chlorella*? Why or why not? Be sure to discuss the rate at both light-limited and light-saturated conditions.

EXPERIMENT 8

Respiratory Control in Potato Tuber Slices and Mitochondria

INTRODUCTION

Potato slices exhibit a low rate of respiration when freshly cut from the tuber, but this rate increases strongly after a period of aerobic incubation ("aging"). The causes for this rise in respiration and the nature of the alternate cyanide-insensitive pathway which becomes active with aging are matters of continuing research interest. It is likely that cutting the tissue in some way damages it; lipids (from membranes) are broken down and the rate of respiration is slow immediately after cutting. After the "aging" period, carbohydrates are metabolized and a higher rate is measured, at least in part as a result of the activity of the alternate, cyanide-insensitive pathway. In a biological perspective, these changes appear to be a part of the transformation from relatively inert storage parenchyma tissue into cells near a cut surface capable of renewing cell divisions leading eventually to wound periderm. Similar events occur when most storage tissues are sliced.

In this experiment, you will measure rates of respiration in freshly cut tissue slices and in slices that have been "aged" overnight. Since oxygen is a substrate of respiration and its concentration in solution may be visualized using the oxygen-sensitive dye Resazurin, rates of oxygen uptake can be determined for these tissues by making spectrophotometric measurements of the incubation solution after several hours. In addition, rates of oxygen uptake in mitochondria (isolated from fresh and aged potato tissue) will be measured using the oxygen electrode. In the case of isolated mitochondria, many of the substrates required for oxidative phosphorylation are stripped away during the isolation process. Succinate (a TCA or Krebs cycle intermediate), Pi and ADP must be supplied to potato mitochondria in order to achieve measurable rates of oxygen uptake.

71

In addition to the control rates, you will measure rates in the presence of an uncoupler (CCCP or DNP) and the respiratory poison KCN, which is known to inhibit cytochrome oxidase. Sensitivity to cyanide is taken as preliminary evidence that cytochrome oxidase is the terminal enzyme of electron transport and that the alternate, cyanide-insensitive, pathway is not active. An increase in respiration in the presence of the uncoupler is a sign of limitation of the rate of respiration by lack of free ADP, i.e. the tissue metabolism is turning over more slowly than the potential of the respiratory system to provide ATP. The mechanism for ATP synthesis and patterns of electron flow are illustrated in Fig. 8-1.

MATERIALS

PART I
Equipment
Spectronic 20
13 x 100 mm disposable test tubes, which
   are usable in the Spectronic 20
test tube racks
Pasteur pipets and bulbs
latex gloves
pipettors

Solutions
Resazurin solution (0.5 mg/100 mL)
50 mM KCN NOTE: toxic; handle with care.
10 mM carbonyl cyanide m-chlorophenyl
    hydrazone (CCCP) in DMSO NOTE: toxic;
    use with care
0.5% Photo-Flo solution
Vaspar: 1:1 petroleum jelly/parafin, warmed

Plant material
Aged and fresh potato (*Solanum tuberosum*) discs, made with a #4 cork borer
Aged and fresh potato discs made with a #4 cork borer and preincubated in KCN

PART II
Equipment
O_2 electrode
recorder
pipettors
aspirator
water bath at 25°C
magnetic stirrer and stir bar

Solutions
Reaction mix: 10 mM Na_2PO_4-KH_2PO_4 buffer (pH
    7.4), 0.3 M mannitol, 10 mM KCl, 5 mM $MgCl_2$
150 mM succinic acid, pH 7.4
15 mM ADP
50 mM KCN NOTE: toxic; use with care
50 mM 2,4-dinitrophenol (DNP) in 95% ethanol
    NOTE: toxic; use with care

Plant material
Mitochondria made from fresh and aged discs. One hundred grams of each tissue were ground in 100 mL homogenization mix and filtered through the cheesecloth. Each extract was sedimented at 1500 g for 20 minutes. The supernatant was decanted and sedimented at 12,000 g for 20 minutes. The supernatant was discarded and the pellets for each were resuspended in 4 mL washing medium and pooled. Each mix was again sedimented at 12,000 g for 20 minutes. The supernatants were discarded and the pellets for each were resuspended in 5 mL of washing medium. Protein content was determined and the protein concentration adjusted to 1 mg/mL.

Waste containers for KCN, DNP and CCCP (**PARTS I** and **II**)

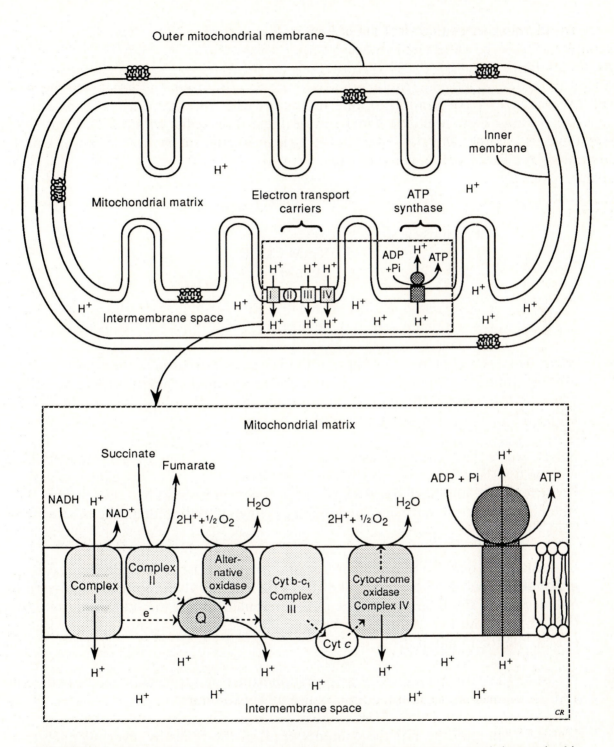

FIGURE 8-1. The mitochondrion and an expanded view of the electron transport chain contained in the inner mitochondrial membrane. Patterns of electron transport are indicated. ATP synthesis is dependent upon the proton gradient formed when electron transport occurs. KCN inhibits cytochrome oxidase activity, but not the alternative oxidase. Q represents the ubiquinone pool. Dotted lines indicate electron flow.

PART I. The Measurement of the Rate of Respiration in Potato Tuber Tissue.

PROCEDURE

(1)   Wash six 13 x 100 mm disposable test tubes with detergent, rinse them with the
        0.5% Photo-Flow solution. Finally, rinse them thoroughly with distilled water and
        invert to drain. The glass must be very clean to prevent the formation of bubbles
        along the inside wall of the glass tube.

(2)   Label the tubes for the treatments listed below.

                1.   Fresh tissue
                2.   Fresh tissue + 20 μL 10 mM CCCP
                3.   Fresh tissue + 0.1 mL 50 mM KCN
                4.   Aged tissue
                5.   Aged tissue + 20 μL 10 mM CCCP
                6.   Aged tissue + 0.1 mL 50 mM KCN

(3)   Wearing latex gloves, transfer 5 mL of the Resazurin solution to the test tubes, then
        add the appropriate amount of inhibitor. **Use automatic pipettors and be careful
        when adding the CCCP and KCN; both are toxic.** Should either solution splash
        onto your skin, wash the area immediately.

(4)   Select twelve discs each of freshly cut tissue and aged tissue. Also select six fresh
        discs and six aged discs which have been preincubated in KCN (wear latex gloves).
        Blot the discs and add six discs of the appropriate tissue to the tube for each
        treatment. Try to remove any small bubbles clinging to the side of the tube with the
        glass rod.

(5)   Use a Pasteur pipet to place about 0.5-1 mL of the warmed Vaspar mixture at the
        upper surface of the Resazurin solution. You want to achieve a good seal with **no
        bubbles** inside the tube. When the plug has sealed, invert the tubes to ensure that
        there are no leaks. Add more Vaspar if necessary. Allow the solutions to
        equilibrate for 15 minutes.

(6)   Measure the absorbance using a spectrophotometer set at 600 nm. Schedule an
        additional measurement of the absorbance after two hours.

(7)   Every fifteen minutes, mix the contents of all of the tubes by inverting each tube.
        When the discs have moved to the sealed end, return the tube to the upright
        position. At the end of one hour and forty-five minutes, invert the tubes for the last
        time, return the tubes to the upright position and allow the solutions to equilibrate

for the final 15 minutes. Check the zero reading and, if necessary, readjust the zero before taking your final readings of the absorbance. Record your results in the chart on page 77. If bubbles are observed inside the tube, regard your results with suspicion.

CALCULATIONS

Determine the change in absorbance (ΔA) for each treatment, by subtracting the reading for each time from the initial absorbance.

PART II. The Measurement of Rates of Mitochondrial Respiration with the Oxygen Electrode. Follow the instructions given in Expt. 7 to calibrate the electrode.

PROCEDURE

(1)   Add 1.5 mL of the reaction mix to the chamber, then add 0.1 mL of the substrate succinic acid. Finally, add 0.4 mL of the mitochondrial preparation made from fresh tissue and let the reaction run until a steady rate can be measured. Record the rate in the chart on page 77.

(2)   Now add 50 μL of 15 mM ADP to the reaction cell and determine the rate of oxygen uptake.

(3)   Add 50 μL of KCN to the reaction cell. Determine the rate of oxygen uptake in the presence of this inhibitor of cytochrome oxidase.

(4)   Rinse out the reaction cell, first with ethyl alcohol, then with distilled water. Repeat step (1), then add 10 μL of 50 mM DNP (an uncoupler) in ethyl alcohol. Determine the rate of O_2 uptake. NOTE: you have assumed that a 2% final concentration of ethyl alcohol has not affected the rate; if time permits, do the proper control to determine if this is the case.

(5)   Repeat all of the above steps, but use the mitochondrial preparation made from aged tissue.

CALCULATIONS

Calculate the rates of oxygen uptake for the six treatments following the instructions given in Experiment 7. Use the chart on page 77.

TEXT REFERENCES

Galston, A. W., P. J. Davies, and R. L. Satter, *The Life of the Green Plant* (3rd ed.), pp. 119-129. Englewood Cliffs, NJ: Prentice Hall, 1980.

Salisbury, F. B. and C. W. Ross, *Plant Physiology* (4th ed.), Chap. 13. Belmont, CA: Wadsworth Publishing Co., 1991.

Taiz, L. and E. Zeiger, *Plant Physiology*, pp. 265-282. Redwood City, CA: Benjamin/Cummings Publishing Co., Inc., 1991.

FURTHER READING

Lance, C., M. Chauveau and P. Dizengrimel, "The cyanide-resistant alternate path in higher plant respiration," pp. 202-247 in R. Douce and D. A. Day, eds., *Encyclopedia of Plant Physiology, New Series*, Vol. 18, *Higher Plant Cell Respiration*. Berlin: Springer-Verlag. 1985.

Laties, G. G. "The cyanide-resistant, alternative path in higher plant respiration," *Annu. Rev. Plant Physiol.* (1982), 33: 519-555.

EXPERIMENT 8

Name _____

Date _____

RESULTS

PART I.

Determine the change in absorbance and determine the rate as $\Delta A/h$.

Time:	0	2 h		Rate
Treatment	A_{600}	A_{600}	ΔA	$(\Delta A/h)$
Fresh				
Fresh + CCCP				
Fresh + KCN				
Aged				
Aged + CCCP				
Aged + KCN				

PART II.

Record the rate for each treatment as units/min, then determine the rates of oxygen uptake.

Treatment	Rate (units/min)	x O_2/unit (μmol/mL/min)	x 60 min/h (μmol/mL/h)
Fresh / Fresh + ADP	/	/	/
Fresh / Fresh + KCN	/	/	/
Fresh / Fresh + DNP	/	/	/
Aged / Aged + ADP	/	/	/
Aged / Aged + KCN	/	/	/
Aged / Aged + DNP	/	/	/

QUESTIONS

1.   Discuss the effect of the uncoupler and KCN on the respiration rate in aged and
      fresh **tissue** (Part I).

2.   Why wasn't the adenylate reductant NADH (normally a source of electrons in
      oxidative phosphorylation) required in the experiments which measured O_2 uptake
      in mitochondria?

3.   How did added ADP and the uncoupler affect the rate of O_2 uptake in
      mitochondria from fresh tissue? Why? What was the effect of added KCN?

4.   Did ADP and the uncoupler affect the rate of O_2 uptake in mitochondria isolated
      from aged tissue? Why might you expect each to have an effect? If they did not,
      give two possible explanations for the results. HINT: one explanation involves the
      changes that occur when potato tissue is aged, the other would be a result of a poor
      mitochondrial preparation. What added information does the addition of KCN
      supply?

5.   Design an experiment to see whether the changes that occur in respiration during
      aerobic incubation ("aging") require active metabolism or requires the synthesis of
      new proteins.

EXPERIMENT 9

Tissue-Water Relations in Potato

INTRODUCTION

Water potential (Ψ) is a measure of the tendency of water to move into or within a system, such as a plant tissue, the soil or the atmosphere. It is the amount of energy per unit volume (or pressure) and is expressed in units of megapascals (MPa) or bars. The relationship between water potential and its components* may be expressed mathematically as:

$$\Psi = \Psi_s + \Psi_p + \Psi_m \qquad \text{Eq. (9-1)}$$

where:

Ψ is the **water potential** of a cell. Plant cells usually have a negative water potential, unless they are in equilibrium with pure water. Such cells are fully turgid and have a Ψ equal to zero.

Ψ_s is the **solute** or **osmotic potential**, the contribution made by dissolved solutes to water potential. It is always negative in sign.

Ψ_p is the **pressure potential**, the contribution made by pressure to Ψ. Cells with turgor pressure have a positive Ψ_p. Cells at incipient plasmolysis have a Ψ_p equal to zero. The development of tension is expressed as a negative Ψ_p, as found in xylem elements in plants undergoing transpiration.

Ψ_m is the **matric potential**, the contribution made by water-binding colloids in the cells, and is negative in sign.

*NOTE: The components of water potential are sometimes identified by different names and symbols. Water potential may be expressed as $\Psi = P - \pi$. In this equation, P (the hydrostatic pressure) is equivalent to Ψ_p; however, π is called the osmotic pressure and is equal to the negative of the solute potential ($\Psi_s = -\pi$).

The object of this experiment is to measure the water potential (Part I) and osmotic potential (Part II) in a plant tissue, then calculate the pressure potential of that tissue. For convenience, the matric potential (Ψ_m), which is often quite small, may be ignored. In fact, some texts consider that matric potential as a component of the solute potential. Thus:

$$\Psi_p = \Psi - \Psi_s \qquad\qquad \text{Eq. (9-2)}$$

MATERIALS

PART I	PART II
Equipment	**Equipment**
No. 5 cork borers	blender
balances and weighing paper	freezing thermometers
150-mL beakers	35 x 300 mm culture tubes for freezing
filter paper or paper towels	magnetic stirrers and oval stirring bars
	potato peelers
Solutions	knives
sucrose solutions: 0.1-0.7 molal	Miracloth
	150-mm funnels
PARTS I and II	Parafilm
Plant material	
potato (*Solanum tuberosum*) tubers or other tuber tissue	

PART I. The Determination of the Water Potential of Potato Tuber Tissue.

When a plant tissue is placed in a solution of sucrose and water, there will be a net movement of water into or out of the tissue (depending on the relative water potentials: water always moves from a less negative Ψ to a more negative Ψ), unless the tissue and solution are at equilibrium. Water movement can be detected by measuring the weight of the tissue before and after an incubation period in the solution. If a series of solutions of different concentrations is used, the concentration of sucrose that would cause no change in weight may be determined by plotting the percent change in weight against sucrose concentration. The sucrose solution which would cause no change in weight may be assumed to have a water potential equal to that of the tissue.

PROCEDURE

(1) Label eight 150-mL beakers for the sucrose solutions listed below and add approximately 75 mL of the correct solution to each beaker:

Distilled water, 0.1, 0.2, 0.3, 0.4, 0.5, 0.6, and 0.7 molal sucrose.

(2)   Use a No. 5 cork borer to cut 16 cylinders from a large potato and trim each cylinder to approximately 4 cm in length, removing any remaining peel. Place the potato cylinders in a covered beaker as you cut them to prevent them from drying out. This step should be completed quickly to prevent excessive evaporation from the cylinders. You may save the remaining potato tissue for Part II if you keep it covered.

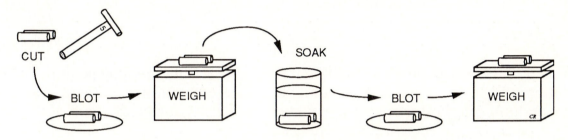

FIGURE 9-1. Procedure for determination of % change in weight.

(3)   Blot (with filter paper or paper towels) and weigh (to 0.01 g) the cylinders in **sets of two** (Fig. 9-1) and record the weight in the chart on page 85. Put one set of cylinders in each of the beakers prepared in step (1).

(4)   After 90 minutes, remove the cylinders, blot with paper towels and re-weigh.

CALCULATIONS

Subtract the initial weights from the final weights. Divide the difference by the initial weight and multiply by 100 to get the percent weight change. Use the table on page 85. Plot the percent change in weight (ordinate) vs. sucrose concentration (abscissa) on the graph on page 85. Determine the molal concentration of sucrose that gives 0% change in weight. Obtain the Ψ_s in bars of that sucrose solution using the formula:

$$\Psi_s = -miRT \qquad\qquad \text{Eq. (9-3)}$$

where:

        m = molality (NOTE: 1 molal = 1 x 10^3 mol m^{-3} H$_2$O)
        i = ionization constant = 1 for sucrose
        R = gas constant = 8.31 J K^{-1} mol^{-1}
        T = room temperature in K ($^\circ$C + 273 = K)

In a free standing solution there is no turgor pressure (Ψ_p = 0), so the Ψ of the solution is equal to the Ψ_s of the solution. At equilibrium, the water potential of the tissue is equal to the water potential of the solution. Convert molality to mol m^{-3}, then use the above equation to determine the solute potential of the solution. Finally, determine the water potential (Ψ) of your potato tissue. Your answer will be in units of J m^{-3} (energy per unit volume) which are equivalent to Pa (pressure). Convert to MPa by dividing by 10^6.

PART II. The Determination of the Solute Potential of Extracted Sap by Cryoscopy.

Cryoscopy is the determination of the freezing point of a liquid and provides a relatively easy means of arriving at the osmotic potential of a solution. Dissolved solutes decrease the freezing point of an aqueous solution. Once the freezing point of a solution or plant extract has been determined, the solute potential may be calculated according to the following relationship, derived from empirical evidence:

$$\Psi_s = 1.22\,\Delta \qquad\qquad \text{Eq. (9-4)}$$

where Ψ_s is the solute potential in MPa at 0 °C and Δ is the freezing point in °C of the solution. Since you want to determine the solute potential for extracted sap at room temperature, you must correct this equation by multiplying by the ratio of the absolute temperatures (room temperature in K/273 K).

PROCEDURE

Sap extraction.

(1)   Remove the peel from two to four potatoes (use leftovers from Part I if possible), quickly chop the potatoes and place them in a blender. Puree the tissue.

(2)   Filter the blended potato mash through Miracloth to remove cell walls and debris. Store the extracted sap in a covered beaker to avoid water loss by evaporation.

Freezing point measurement.

(1)   Check the zero point of the Heidenhain thermometer by immersing it in a crushed ice-distilled water mixture. Record the reading in the space on page 86. Your final calculations will be corrected to this zero point.

(2)   Place a 60-mL sap sample in a large culture tube along with a magnetic stirring bar and insert the thermometer.

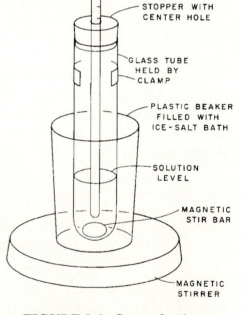

FIGURE 9-2. Set-up for the freezing point determination.

(3)   Constant stirring of the sap is essential. Place the tube in position on a magnetic stirrer and make sure that the solution is being stirred vigorously (Fig. 9-2).

(4)   Surround the freezing tube with an ice-salt bath which is at about -8 ° to -10 ° C.

(5)   When the thermometer reads about + 1 ° C, the temperature should be recorded every 10 seconds (use the chart on page 86).

Watch the thermometer carefully. There should be a continuous drop in temperature and, if supercooling does not occur, a plateau may be observed for a short time at the freezing point. If supercooling does occur, there will be a rapid drop in temperature, followed by an increase to a plateau which represents the **apparent** freezing point. In this case, be sure to record the lowest temperature reached as well as the apparent freezing point. See Fig. 9-3.

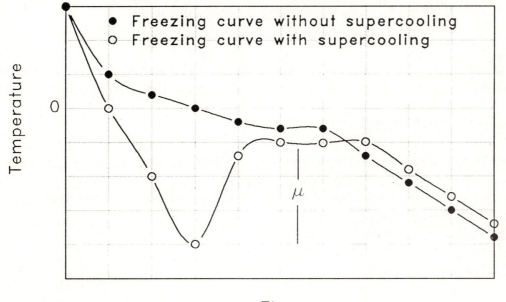

FIGURE 9-3. Sample freezing curves with and without supercooling.

Usually the water does not freeze at the true freezing point but undergoes supercooling. When the temperature of a solution reaches the actual freezing point, the molecules of solution may not be correctly aligned for crystallization; the temperature may continue to drop without the onset of freezing. Eventually, some pure water crystallizes out of the sap solution, causing the remaining solution to be more concentrated. As soon as the freezing of water begins, its crystalline structure expands, causing heat to be liberated, and the mercury rises rapidly. The water remaining in solution freezes at a lower temperature (the apparent freezing point) than the true freezing point of the solution since the sap is now more concentrated.

CALCULATIONS

Plot temperature against time on the graph on page 87.

To determine the **true freezing point** (Δ), you must make a correction for supercooling, defined as the difference between the apparent freezing point (Δ') and the temperature at which crystallization of water began. Use the formula:

$$\Delta = \Delta' - 0.0125\,\mu \qquad\qquad\qquad \text{Eq. (9-5)}$$

where:

Δ = true freezing point
Δ' = apparent freezing point
μ = degrees of supercooling (negative in sign)
0.0125 = amount of water (1/80) solidifying
per degree of supercooling

When the true freezing point (Δ) has been calculated, make another correction for the zero point of the thermometer. Then use Eq. (9-4) to determine the solute potential (Ψ_s).

$$\Psi_s = 1.22\,\Delta \quad \text{at } 0\,^\circ C$$

Remember to correct to room temperature by multiplying by the ratio of absolute temperatures. Record the Ψ_s in the space provided on page 88. Assume your determined Ψ_s is equal to the osmotic potential of the sap in the intact tissue. Calculate the pressure potential (Ψ_p) for the cells in your tissue.

TEXT REFERENCES

Galston, A. W., P. J. Davies, and R. L. Satter, *The Life of the Green Plant* (3rd ed.), pp. 137-145. Englewood Cliffs, NJ: Prentice Hall, 1980.

Salisbury, F. B. and C. W. Ross, *Plant Physiology* (4th ed.), pp. 44-65. Belmont, CA: Wadsworth Publishing Co., 1991.

Taiz, L. and E. Zeiger, *Plant Physiology*, pp. 61-79. Redwood City, CA: Benjamin/Cummings Publ. Co., Inc., 1991.

FURTHER READING

Dainty, J., "Water relations of plant cells," pp. 12-35 in U. Lüttge and M. G. Pitman, eds., *Encyclopedia of Plant Physiology, New Series*, Vol. 2, *Transport of Plants II, Part A, Cells*. Berlin: Springer-Verlag, 1976.

Kramer, P. J., *Water Relations of Plants*. Santa Clara, CA: Academic Press, 1983.

EXPERIMENT 9

Name _____

Date _____

RESULTS

PART I.

Record the initial and final weights and determine the %Δ weight.

Molal [Sucrose]	0	0.1	0.2	0.3	0.4	0.5	0.6	0.7
Initial Weight (g)								
Final Weight (g)								
Δ Weight (g)								
% Δ Weight								

Plot a graph below of the percent weight change vs. sucrose concentration. Draw the best-fit straight line through the points.

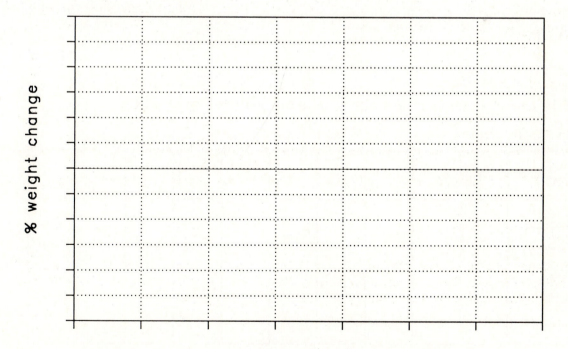

% weight change

sucrose concentration (molal)

Sucrose concentration at which zero percent change in weight is observed: _____

Determine the Ψ_s of the sucrose solution using Eq. (9-3). Show your calculations.

Now determine the Ψ of the potato tissue (see page 81). Show your calculations.

Ψ of the potato tissue = _____

PART II.

Thermometer reading in ice and distilled water mix: _____

Record the change in temperature with time in the chart below.

Time:											
Temperature:											
Time:											
Temperature:											

Plot a graph of temperature vs. time below.

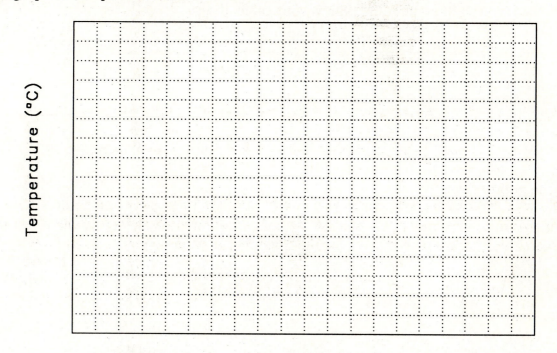

Time

Determine the true freezing point of your extracted sap using Eq. (9-5). Show your calculations.

True freezing point: _____

Correct this figure for the zero point on your thermometer:

Corrected freezing point: _____

Finally, calculate the Ψ_s of the extracted potato sap using Eq. (9-4). Be sure to correct your answer to room temperature. Show your calculations.

Ψ_s of the potato tissue = _____

Assume your determined Ψ_s is equal to the solute potential of the sap in the intact tissue. Calculate the pressure potential Ψ_p for the cells in your tissue with Eq. (9-2).

Ψ_p of the potato tissue = _____

QUESTIONS

1.   Which point on the graph from Part I represents a solution whose water potential was the same as that of the potato tissue at the beginning of the experiment? Explain.

2.   Which tissues have a water potential equal to that of their sucrose solution at the end of the experiment?

3.   What deficiencies are present in the experimental design? What assumptions were made which might affect your results?

EXPERIMENT 10

The Pressure Bomb and Determination of Water Potential

INTRODUCTION

Plant water status is usually expressed as water potential (see Expt. 9). Unfortunately, it is difficult to measure the water potential directly, although several indirect methods are available. All have shortcomings, so for each application there is usually one method which is most suitable. These methods are described in most recent plant physiology textbooks.

In this experiment, you will measure the water status of *Xanthium* plants which have been subjected to soil drying (water was withheld for six or seven days) and those experiencing transpiration in wet soil. You will learn to measure water potential using the pressure bomb (Fig. 10-1).

The water potential (Ψ) of a leaf at equilibrium is the same throughout the leaf, although the contributions of the various components will differ in different cells and tissues. In the xylem, the main contribution is a negative turgor pressure (tension). The xylem sap can be assumed to be essentially pure water (or, at least, to have very few dissolved solutes), so that you can consider

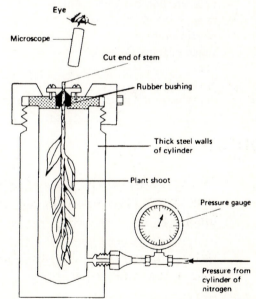

FIGURE 10-1. The pressure bomb. (Adapted from Scholander *et al.,* 1966, *Plant Physiol. 41*:529-532.)

the xylem turgor to be exactly equal to its water potential: $\Psi = \Psi_s + \Psi_p$, but in xylem elements $\Psi_s = 0$, so $\Psi = \Psi_p$.

When a leaf is cut from the plant, the tension in the xylem is released; Ψ_p becomes equal to zero and, consequently, Ψ becomes equal to zero. The xylem sap recedes from the cut surface as water moves into the surrounding cells (where the Ψ is more negative) along a Ψ gradient. The leaf is then placed in the pressure bomb where just enough pressure is applied to force the xylem sap back to the cut surface. The amount of pressure is equal but opposite in sign to the tension which existed in the xylem at the time the leaf was cut. In other words, the applied pressure is equal to $-\Psi_p$.

Since $\Psi_p = \Psi$ in the xylem elements and you may consider the xylem cells to be in equilibrium with the surrounding leaf cells, the applied pressure is also equal to $-\Psi_{leaf}$, giving a measure of the leaf water potential.

MATERIALS

Equipment
pressure bomb with N_2 source
magnifying glass
plastic wrap
single edge razor blades
lamp
Kimwipes
manufacturer's instructions for the pressure bomb

Plant material
Cocklebur (*Xanthium strumarium*)
plants, 4 weeks old, some
deprived of water for 4 days

PROCEDURE

(1)   Choose a *Xanthium* plant that has been well watered. Each plant should have four or five fully expanded leaves. You may practice on as many of these as you need, but make your recorded measurement of Ψ on the youngest mature leaf. Ask the instructor if you are not sure which leaf to use. Wrap the leaf in plastic wrap to prevent water loss by transpiration or evaporation and cut the leaf at the base of a petiole with a razor blade. Make a **single** cut, perpendicular to the axis of the petiole.

(2)   Seal the leaf into the pressure chamber. **IMPORTANT: Be certain the lid is properly sealed before applying any pressure.**

(3)   Carefully following the printed instructions for the pressure bomb, increase the pressure in the chamber **slowly** (about 0.3 bars/s), while observing the cut end of the petiole through a magnifying glass. When the cut end is just wetted, turn off the

pressure and record the applied pressure in the chart on page 93. Most pressure bombs are calibrated in bars, so you must convert to MPa by dividing by 10.

Sometimes a bubble will come to the cut end before the water column and look like an end point. These false endpoints can be distinguished because drying the end with a Kimwipe will leave the end dry, whereas a true endpoint will rewet. If **many** bubbles appear, you may have missed the endpoint.

(4)   Turn the pressure knob to exhaust and **wait until all of the pressure is released**. Remove your leaf from the pressure bomb.

(5)   Repeat steps (1) through (4) with a leaf from a water stressed plant.

CALCULATIONS

Consider the equations in the introduction and report the Ψ of the leaves you tested in the chart on page 93.

TEXT REFERENCES

Galston, A. W., P. J. Davies, and R. L. Satter, *The Life of the Green Plant* (3rd ed.), pp. 159-160. Englewood Cliffs, NJ: Prentice Hall, 1980.

Salisbury, F. B. and C. W. Ross, *Plant Physiology* (4th ed.), pp. 59-60. Belmont, CA: Wadsworth Publishing Co., 1991.

Taiz, L. and E. Zeiger, *Plant Physiology*, pp. 75-6. Redwood City, CA: Benjamin/Cummings Publishing Co., Inc., 1991.

FURTHER READING

Scholander, P. F., H. T. Hammel, E. D. Bradstreet and E. A. Hemmingsen, "Sap pressure in vascular plants," *Science* (1965), 148:339-346.

Tyree, M. T. and H. T. Hammel, "The measurement of the turgor pressure and the water relations of plants by the pressure-bomb technique," *Journal of Experimental Botany* (1972), 23(74):267-282.

Waring, R. H. and B. D. Cleary, "Plant moisture stress: Evaluation by the pressure bomb," *Science* (1967), 155:1248-1254.

EXPERIMENT 10   Name _____

                                                  Date _____

RESULTS

Record the applied pressure in the chart below and determine the water potential (in MPa) of the leaves you measured.

Treatment	Applied Pressure (bars)	Water Potential (MPa)
Watered		
Water Stressed		

QUESTIONS

1.   When you remove the leaf from the plant, why is it important **not** to recut the petiole to get a better end?

2.   In what situation would the pressure bomb be **in**appropriate for measuring Ψ? What method could you use in that case?

3.   What effect did the water stress treatment have on Ψ? Why?

4.   What are the main sources of error in this experiment? Are the assumptions you made valid? How might they affect your determination of the water potential?

EXPERIMENT 11

Transpiration and the Mechanism of Guard Cell Movement

INTRODUCTION

Transpiration may be defined as the loss of water vapor from the shoots of a terrestrial plant. When transpiration rates are high, this process provides all the energy required to move water and associated minerals through the plant; the difference between the water potential of the soil and the water potential of the air is the driving force for this movement of water. The transpiration rate (Δ weight/Δ time), then, is determined by this driving force and by the resistance of the plant, the soil, and the boundary layer of air to water flow. For a steady state situation:

$$\frac{\Delta \text{weight}}{\Delta \text{time}} = \frac{\Psi_{soil} - \Psi_{air}}{r_{soil} + r_{plant} + r_{air}} \qquad \text{Eq. (11-1)}$$

where:

$$
\begin{aligned}
\Psi_{soil} &= \text{the water potential of the soil} \\
\Psi_{air} &= \text{the water potential of the air} \\
r_{soil} &= \text{the resistance of the soil to water movement} \\
r_{plant} &= \text{the resistance of the plant to water movement;} \\
&\quad \text{the sum of } r_{roots}, r_{xylem} \text{ and } r_{leaves} \\
r_{air} &= \text{the resistance of the air (particularly the boundary} \\
&\quad \text{layer) to water movement}
\end{aligned}
$$

This is not a simple linear function since the resistances of the air and soil to water flow are not constant. These complexities make a rigorous quantitative description of transpiration difficult, but you will still find the concepts of a driving force and resistance useful in explaining the results you will obtain in this experiment.

95

The water potential of the air is a logarithmic function of the relative humidity (RH):

$$\Psi_{air} = \frac{RT \ln (\%RH/100)}{V_w} \qquad \text{Eq. (11-2)}$$

where:

$R = $ the gas constant, 8.31 J K^{-1} mol^{-1}
$T = $ temperature (K)
$V_w = $ partial molal volume of water, 18.03 m^3 mol^{-1}

Note that Ψ_{air} is then calculated as J m^{-3} (energy per unit volume = pressure as Pa). You may convert to MPa by dividing by 10^6.

The relative humidity of the air may be readily measured with a sling psychrometer and the Ψ_{air} determined using the equation above. By measuring the rate of water loss from the plant at different relative humidities, you can determine the effect of changes in Ψ_{air} on the rate of transpiration as suggested by Eq. (11-1).

Your plants will be placed in water during the experiment, so Ψ_{soil} and r_{soil} will be equal to zero. Under some conditions, r_{air} (the resistance of the boundary layer of air over the leaf) can be important, but will not be measured in your experiment. The resistance of the plant is the sum of the resistance of the roots, xylem and leaves. The resistance of the xylem is nearly zero and the resistance of the roots is **usually** negligible in comparison to the leaves (the influence of the r_{roots} can be explored by carrying out the experiment in ice water). For the purposes of this experiment, r_{plant} may be considered equal to r_{leaves}.

The resistance of the leaf to water loss is a consequence of its structure. Terrestrial plants are covered with a waxy cuticle which has a very high resistance to water flow. Open stomata provide an alternative low resistance pathway through which water vapor can move, but closed stomata offer relatively high resistance. The resistance of the leaves, then, may be considered to be equal to the resistance of the stomata (r_s).

Environmental factors influence the components of Eq. (11-1) and hence can change the transpiration rate in three distinct ways:

1.   As physical factors which determine the value of the driving force.
2.   As physical factors which influence the resistance of the soil, the plant or boundary layer of air over the leaf.
3.   As stimuli which influence the movements of the guard cells and thus change the resistance of the leaf.

In this experiment you will alter the relative humidity (factor 1) and the resistance of the leaf (factor 3). The plants provided have been kept in the dark in aerated water until the start of the experiment. The stomata should be closed and no water deficit should be present, much as would exist in a plant at dawn. The Ψ of the leaf cells should be close to zero.

As the transpiration rate increases through the day (especially on a hot, dry summer day) a water deficit may build up in the leaves; that is, the Ψ of the leaves may become more negative. Your experiment will attempt to duplicate the daily changes in environment within a few hours time. Since the water potential may be influenced by these environmental changes, you will measure the Ψ of the primary leaves of your plants at the beginning and at the end of the experiment using the pressure bomb.

As should be clear from the above discussion, stomata play an important role in determining rates of transpiration by changing the resistance of the plant. In Part II of this experiment, you will investigate some internal and external factors which influence stomatal opening. Epidermal strips of *Vicia* will be floated on experimental solutions in the light and dark to test for the requirements for K^+ and light in stomatal opening.

MATERIALS

PART I
Equipment
humidity tents and humidifiers
balances (torsion or other)
lamps
Parafilm
round bottom flasks
125-mL flasks

sling psychrometers and instructions
pressure bomb
hair dryer
Saran wrap
razor blades

Plant material
Cocklebur (*Xanthium strumarium*) plants, 4 weeks old, kept in the dark for 24 h.

PART II
Equipment
razor blades
slides and cover slips
Pasteur pipets and bulbs
beakers
microscopes

Solutions
100 mM KCl in 2.5 mM MOPS
200 mM mannitol in MOPS
100 mM choline chloride in 2.5 mM MOPS
100 mM KCl + 10^{-5} M abscisic acid (ABA)
   in 2.5 mM MOPS

Plant material
Broad bean (*Vicia faba* var. Broad Windsor) plants, 6 weeks old, kept in the dark for 24 h.

PART I. The Measurement of Transpiration in Whole Plants.

PROCEDURE

(1)   Choose a *Xanthium* plant that has been stored in the dark for the last 24 hours, remove the soil from around the roots and place the plant in a 125-mL Erlenmeyer flask with about 100 mL of distilled water. Be sure that there is no water on the leaves of the plant. Remove one of the lower leaves, choosing one that has a leaf opposite it on the stem. Measure the Ψ of this leaf using the pressure bomb (see Expt. 10) and record it in the chart on page 103.

(2)   Place the plant in the flask and seal around the stem with a piece of Parafilm. Place the plant on a balance in the humidity tent. If a torsion balance is used, place the plant on one pan of the torsion balance, a sealed flask of water weighing several grams less on the other pan and balance the two flasks by adding weight to the pan containing the flask of water. Record the initial balancing weight and all subsequent weight measurements in the chart on page 103.

(3)   Measure the change in weight of the plant at 10 minute intervals for 40 minutes. Be sure to measure the relative humidity (% RH) inside the tent using a sling psychrometer once during this period. Whirl the body of the psychrometer inside the humidity tent for one and a half minutes, then read the temperature of the wet and dry bulbs. Insert the body of the psychrometer inside the tube and line up the wet and dry bulb temperatures to read the %RH. More specific instructions for the use of the psychrometer will be provided by the instructor. Record the %RH on page 103.

(4)   Turn on the light provided and focus it on the plant using a round-bottom flask containing water as a lens. Make measurements of the weight every 10 minutes for 40 minutes. Measure the % RH inside the tent.

(5)   Remove the tent from around the balance and plant and turn off the humidifier. Dry the air around the plant with a hair dryer. With the plant still in the light, measure the weight of the plant every 10 minutes for 40 minutes. Again measure the %RH. At the end of the experiment, remove the leaf **opposite** the leaf previously removed and determine its Ψ using the pressure bomb.

CALCULATIONS

        Determine the cumulative change in weight at each time interval (weight at 0 time minus the weight at each time). Use the chart on page 103. Plot the cumulative change in weight against time on the graph on page 104.

PART II. Investigation of the Mechanism of Guard Cell Movement.

Stomata respond to light by opening. The opening may be triggered directly by blue light; the H^+-ATPase at the plasma membrane is switched on, causing a change in the membrane potential of the guard cells, which leads to the influx of K^+ ions. When potassium ions are taken up by the guard cells, an influx of H_2O follows; the resulting increase in turgor leads to stomatal opening (Fig. 11-1).

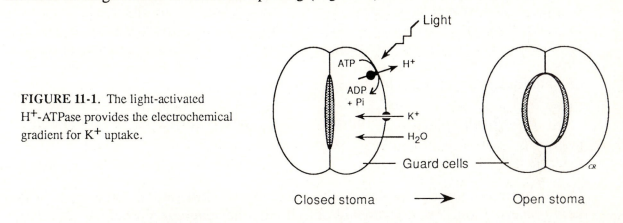

FIGURE 11-1. The light-activated H^+-ATPase provides the electrochemical gradient for K^+ uptake.

Red light may also stimulate opening of the stomata by initiating photosynthesis in the guard cells. The subsequent drop in the local concentration of CO_2 stimulates stomatal opening by enhancing K^+ uptake. The mechanism for red-light stimulated opening of stomata is not well understood.

In this section of the experiment you will perform some of the tests which led investigators to propose that K^+ is required for the movement of guard cells. Epidermal strips of *Vicia faba* containing guard cells will be floated on experimental solutions to test for this requirement for K^+. The effect of abscisic acid (ABA), a plant hormone known to influence stomatal movements, will also be tested.

PROCEDURE

(1)   Prepare five beakers by adding approximately 50 mL of the appropriate solutions for the following treatments:

        1.   100 mM KCl, light
        2.   100 mM KCl, dark
        3.   200 mM mannitol, light
        4.   100 mM ChCl (choline chloride), light
        5.   100 mM KCl, 10^{-5} M ABA, light

(2)   Remove a healthy leaf from the *Vicia faba* plant that has been kept in the dark for 24 hours. Tear the leaf almost parallel to the mid vein, pulling toward yourself with

one hand and away with the other. Epidermal strips should exist along the edge of both torn pieces. Trim the strips with a sharp razor blade; they tangle less if a section of leaf is retained along the edge of the strip. **Do not allow the epidermal strips to dry out**; place several strips directly into each of the solutions above and make sure they are covered with liquid.

(3)   Place the beakers in bright light, except for 2 (the "dark" treatment), which should be placed in a drawer or cabinet. Wait at least 1 hour, then observe the strips at random under the microscope and estimate stomatal apertures. Classify the aperture widths of the stomata in each strip as wide open, open, nearly closed, or closed. This is best done by counting and classifying all the stomata in a given area of epidermal strip, as in the following example:

wide open	open	nearly closed	closed
48	14	12	6

In this example, there were 48 stomata which were classified as wide open in the field of view. Often the stomata will not **all** be either open or closed. By classifying individual stomata, you can determine whether a given condition tends to open or close the stomata. This example was clearly provided with the correct conditions for opening. Record your counts in the chart on page 104 and decide what effect each treatment had on the opening of the stomata.

PART III. Demonstration: The Histochemical Localization of K^+.

The presence of large amounts of K^+ in the guard cells of open stomata and the relative absence of K^+ in the guard cells of closed stomata can be confirmed using a histochemical technique. The method of K^+ localization takes advantage of the fact that sodium cobaltinitrite is soluble, while potassium cobaltinitrite is insoluble. When tissues are placed in a concentrated solution of the sodium salt in 10% acetic acid at 0 °C the cobaltinitrite enters the cells and forms the insoluble potassium salt crystals with the intracellular potassium. The potassium salt, which is yellow, is then reacted with ammonium sulfide to form the sulfide of cobalt, which is black and readily visible under a microscope.

Some epidermal strips from *Vicia faba* were placed in 100 mM KCl. One set was placed in the light; another was placed in the dark. At the end of an hour both were treated with the chemicals described above. Examine the demonstration slides and note **which cells contain the black precipitate** in the epidermal strips which were exposed to light versus those which were kept in the dark. The staining procedure kills the cells so the stomata which were open in the light will appear closed. Draw a representative set of guard cells for each treatment (in the space provided on page 105) and note which cells contain the black precipitate.

TEXT REFERENCES

Galston, A. W., P. J. Davies, and R. L. Satter, *The Life of the Green Plant* (3rd ed.), pp. 147-160. Englewood Cliffs, NJ: Prentice Hall, 1980.

Salisbury, F. B. and C. W. Ross, *Plant Physiology* (4th ed.), pp. 66-87. Belmont, CA: Wadsworth Publishing Co., 1991.

Taiz, L. and E. Zeiger, *Plant Physiology*, pp. 81-98 and 133-6. Redwood City, CA: Benjamin/Cummings Publishing Co., Inc., 1991.

FURTHER READING

Assman, S. M., L. Simoncini, and J. I. Schroeder," Blue light activates electrogenic ion pumping in guard cell protoplasts of *Vicia faba*," *Nature* (1985), 318:285-287.

Kramer, P. J., "Changing concepts regarding plant water relations," *Plant Cell and Environment* (1988), 11:565-568.

Mansfield, T. A., A. M. Hetherington, C. J. Atkinson, "Some current aspects of stomatal physiology," *Annu. Rev. Plant Physiology. Plant Mol. Bio.* (1990), 41:55-75.

Outlaw, W. H., Jr., "Current concepts on the role of potassium in stomatal movements," *Physiologia Plantarum* (1983), 59:302-311.

Outlaw, W. H., Jr., "Critical examination of the quantitative evidence for and against photosynthetic CO_2 fixation by guard cells," *Physiologia Plantarum* (1989), 77:275-281.

Pierce, M. and K. Raschke, "Correlation between loss of turgor and accumulation of abscisic acid in detached leaves," *Planta* (1980), 148:174-182.

Schulze, E-D., E. Steudle, T. Gollan, and U. Schurr, "Response to Dr. P.J. Kramer's article 'Changing concepts regarding plant water relations'," *Plant Cell and Environment* (1988), 11:573-576.

Zeiger, E., "The biology of stomatal guard cells," *Annu. Rev. Plant Physiol.* (1983), 34:441-475.

Zeiger, E., G. O. Farquhar, and I. R. Cowan, eds., *Stomatal Function*. Stanford, CA: Stanford University Press, 1987.

EXPERIMENT 11

Name _____

Date _____

RESULTS

PART I. Record the weight at each time interval. Indicate the % RH during each 40 minute period and the initial and final Ψ of the leaves.

Treatment	Time	Weight (g)	Cumulative Δ Weight (g)	Rate (g/h)
High Humidity, Dark			0	
% RH = _____				
High Humidity, Light			0	
% RH = _____				
Low Humidity, Light			0	
% RH = _____				

Initial pressure bomb reading:

_____

Initial Ψ (MPa):

_____

Final pressure bomb reading:

_____

Final Ψ (MPa):

_____

Plot the cumulative change in weight against time on the graph below. Determine the rate of weight loss per hour (g/h) for each treatment and record it in the chart on the previous page.

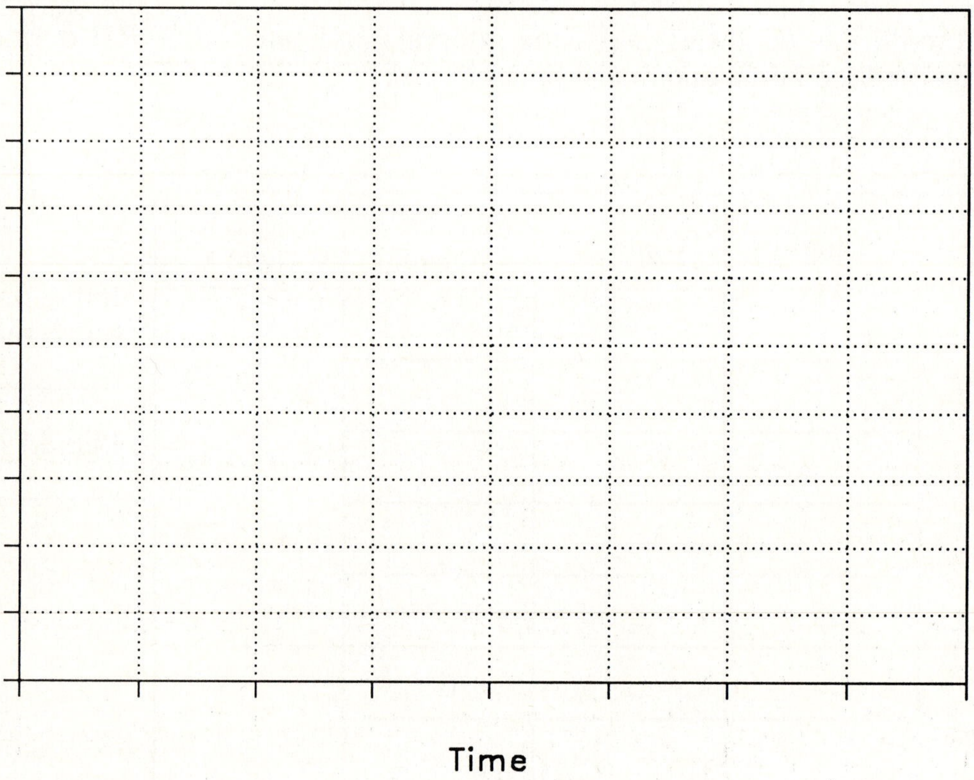

Time

PART II. Present your observations for Part II in the chart below.

#	Treatment	Wide Open	Open	Nearly Closed	Closed	Treatment Causes:
1	KCl, light					
2	KCl, dark					
3	Mannitol,light					
4	ChCl, light					
5	KCl, ABA, light					

PART III. Record your observations of the demonstration slides in the space below. Draw the guard cells and show the location of the black precipitate.

Light	Dark

QUESTIONS

1. How does light affect the transpiration rate? Why? Refer specifically to Eq. (11-1) and to your data.

2. How does a change in the relative humidity affect the transpiration rate? Why? Refer specifically to Eq. (11-1) and to your data.

3. Other environmental factors affecting the rate of transpiration (but not investigated in this experiment) might be changes in temperature, CO_2 concentration, and wind. Briefly state one effect each of these might have on the transpiration rate and why.

4. How was the Ψ_{leaf} affected by changes in the lab environment? Give an explanation for your results.

5. State the reasoning behind each experimental treatment in Part II.

6. What conditions are needed for stomatal opening and why? Refer specifically to your results from Part II and the demonstration slides (if available).

EXPERIMENT 12

Mineral Nutrition

INTRODUCTION

A mineral is defined as essential if normal growth and reproduction cannot be carried out in its absence. Plants usually respond to the deficiency of an essential mineral by characteristic visual symptoms and stunted growth. Such symptoms are of interest because they shed light on the necessary function of the element in the plant and because they are used by agriculturists to determine how and when to fertilize their crops. A trained observer can predict the mineral deficiency of a given soil simply by examining the conditions of the plants growing in it.

In this experiment, you will record the development of symptoms in sunflower and corn growing under several mineral deficient conditions with coded labels. At the end of three weeks, you will attempt to determine which deficiency exists in each set of plants.

MATERIALS

Equipment
1000-mL brown bottles
aluminum weighing pans
non-absorbent cotton
aeration system
greenhouse
automatic pipets

Solutions
see Table 12-1

Plant material
sunflower (*Helianthus annuus* var. Sungold) plants, 3 weeks old
corn (*Zea mays* var. Golden Cross Bantum) plants, 3 weeks old

PROCEDURE

The roots of three healthy sand-grown sunflower seedlings (corn seedlings may be used as well to observe symptoms in a monocot) were rinsed and placed into one of ten containers, each of which was filled with one of the aerated experimental solutions (see Table 12-1). Only the distilled water and the complete medium treatments are labeled. The other eight treatments (-Ca, -S, -Mg, -K, -N, -P, -Fe, and -microelements) are coded.

For the next three weeks, make weekly measurements of the plants' heights and observations of their general appearance. A list of typical deficiency symptoms is included for your guidance (Table 12-2). Symptoms to look for include:

1) Yellow leaves and their location.
2) Dead spots and their location.
3) Peculiarities of venation.
4) Unusual colors.
5) Other pertinent information.

Record your observations in the chart on page 111.

TABLE 12-1. Solutions were made by filling 500-mL brown bottles half-full with distilled water, then adding the proper amount (in mL) of each component (for each treatment read down the column). The solutions were mixed and the bottles were filled to the top with distilled water.

Stock Solution	Stock Solution to Add for Each Treatment (mL)									
	Complete	Ca	S	Mg	K	N	P	Fe	Micro	H_2O
0.5 M $Ca(NO_3)_2$	5	0	5	5	5	0	5	5	5	0
0.5 M KNO_3	5	5	5	5	0	0	5	5	5	0
0.5 M $MgSO_4$	2	2	0	0	2	2	2	2	2	0
0.5 M KH_2PO_4	1	1	1	1	0	1	0	1	1	0
FeNaEDTA	1	1	1	1	1	1	1	0	1	0
Microelements	1	1	1	1	1	1	1	1	0	0
0.5 M $NaNO_3$	0	5	0	0	5	0	0	0	0	0
0.5 M $MgCl_2$	0	0	2	0	0	0	0	0	0	0
0.5 M Na_2SO_4	0	0	0	2	0	0	0	0	0	0
0.5 M NaH_2PO_4	0	0	0	0	1	0	0	0	0	0
0.5 M $CaCl_4$	0	0	0	0	0	5	0	0	0	0
0.5 M KCl	0	0	0	0	0	5	1	0	0	0

TABLE 12-2. The symptoms of mineral deficiency in greenhouse-grown sunflower plants (the symptoms may vary with the species and the growing conditions).

Mineral	Deficiency symptoms in sunflower
Calcium	The first symptom is usually a deformation of the younger leaves, then a **disintegration of terminal growing areas**. Symptoms appear almost **immediately** and there is very **little growth**.
Iron	**Young (upper) leaves rapidly turn light green or almost white**, while older leaves are green. Unlike most other elements, iron cannot be withdrawn from the older leaves, so they retain a normal appearance. The yellowing or chlorosis of the younger leaves is most obvious in the intervenal areas and appears in a very short time.
Magnesium	The older (lower) leaves show symptoms first, yellowing from the tip and eventually falling. **Veins remain green longer** than the intervenal areas.
Nitrogen	The **cotyledons and lower leaves very rapidly turn yellow to brown** and die. There is **very little growth**. In some species like corn, a red or purpling occurs along the veins.
Phosphorous	The plants show little growth, **have slender stems** and have **dark green leaves**. Dead areas develop on leaves, petioles, and (in older plants) fruits, causing some fruit drop. In some plants like corn, the production of anthocyanins may cause a red or purple color.
Potassium	Plants are **stunted** and exhibit a **yellow mottling of the leaves** which **curl and crinkle**. Necrotic areas are seen on the leaf tips and edges and the leaves eventually show an overall yellowing.
Sulfur	In the early stages, there may be some yellowing of the younger leaves; **ultimately an over-all pale green** may dominate. **Growth continues normally**. Eventually the apical leaves appear quite yellow to white.
Microelements	All of the microelements are missing in the unknown. Usually, a **delayed response** is seen with symptoms (like wilting) most evident in the youngest leaves. Typical symptoms for individual microelements follow. Manganese: Leaves become spotted with dead areas and fall. Copper: Tips of young leaves wither; plant may wilt even when it is watered. Zinc: Yellowing of lower leaves at tips and margins; leaves deformed. Molybdenum: Required for nitrogen metabolism, so symptoms resemble those of nitrogen deficiency.

TEXT REFERENCES

Epstein, E., *Mineral Nutrition of Plants*. New York: John Wiley & Sons, Ltd., 1972.

Galston, A. W., P. J. Davies, and R. L. Satter, *The Life of the Green Plant* (3rd ed.), Chap. 7. Englewood Cliffs, NJ: Prentice Hall, 1980.

Salisbury, F. B. and C. W. Ross, *Plant Physiology* (4th ed.), Chap. 5. Belmont, CA: Wadsworth Publishing Co., 1991.

Taiz, L. and E. Zeiger, *Plant Physiology*, pp. 100-119. Redwood City, CA: Benjamin/Cummings Publishing Co., Inc., 1991.

FURTHER READING

Clark, R. B., "Nutrient solution growth of sorghum and corn in mineral nutrition studies," *Journal of Plant Nutrition* (1982), 5:1039-1057.
Glass, A. D. M., *Plant Nutrition. An Introduction to Current Concepts*. Boston, MA: Jones and Bartlett, 1989.
Jones, J. B., Jr., "Hydroponics: Its history and use in plant nutrition studies," *Journal of Plant Nutrition* (1982), 5:1003-1030.
Läuchli, A. and R. L. Bieleski, eds., *Encyclopedia of Plant Physiology, New Series,* Vols. 15A-15B, *Inorganic Plant Nutrition*. Berlin: Springer-Verlag, 1983.
Marshner, H., *Mineral Nutrition of Higher Plants*. London: Academic Press, 1986.

EXPERIMENT 12

Name _____

Date _____

RESULTS

Record your measurements and observations each week in the charts that follow. At the end of three weeks, predict which of the coded bottles contains which solution.

Plant: _____

Date:				Mineral?
Complete				
A				
B				
C				
D				
E				
F				
G				
H				
distilled water				

Plant: _____

Date:				Mineral?
Complete				
A				
B				
C				
D				
E				
F				
G				
H				
distilled water				

QUESTIONS

1.   Indicate what symptoms led you to make your choices. Be specific.

2.   Does the presence of a symptom on young leaves or older leaves give you any
      information as to the mobility of a mineral? Cite specific examples from your data
      to support your answer.

EXPERIMENT 13

Nitrate Reductase: The Transformation of *Chlamydomonas*

INTRODUCTION

Most of the nitrogen required for plant growth is obtained from the soil as nitrate (NO_3^-) ions. The nitrate must be converted to a more reduced form before it is usable in plant metabolism. This conversion is carried out by the enzyme nitrate reductase, which, in conjunction with NAD(P)H, reduces the nitrate ions to nitrite ions (NO_2^-), which are subsequently reduced to ammonium (NH_4^+) by another enzyme, nitrite reductase. A plant which lacks the essential enzyme nitrate reductase must be supplied with nitrogen in a more reduced form than nitrate.

Chlamydomonas reinhardtii, a unicellular green alga, uses nitrate reductase to reduce the nitrate ions it absorbs from the surrounding medium. A mutant strain of *Chlamydomonas*, nit1-305, lacks this essential enzyme (the gene is defective) and will not grow unless the supplied medium contains ammonia. In order to show that the nitrate reductase gene is required for the reduction of nitrate, the mutant strain may be rescued by transforming it with wild-type DNA which codes for nitrate reductase. Transformation is the incorporation of new DNA into the cell and the expression of the new gene. Individual transformed cells will grow and divide on standard medium (lacking ammonia), producing colonies, but those cells which do not incorporate the new DNA will not survive.

Genetic transformation, in the case of *Chlamydomonas*, is a simple procedure. The nitrate reductase-deficient mutant you will use is a double mutant which also lacks a cell wall (nit1-305 cw-15); the absence of the cell wall makes the process of transformation more efficient. The gene which encodes for nitrate reductase is contained in a 14.5 kilobase fragment of *Chlamydomonas* DNA which has been incorporated into a circular

115

piece of DNA or plasmid (pMN24). When glass beads, polyethylene glycol, and the plasmid DNA are added to the mutant cells, and the cells are agitated in a vortex mixer, the plasmid (containing the desired gene) may enter the cell and be incorporated into the nuclear genome. You may test for incorporation of the gene by growing the cells on medium lacking ammonia. Colonies of transformed cells will appear in one to two weeks.

Once the transformed colonies have grown on agar medium, they will be transferred to liquid SGII-NO$_3$ medium and placed on a shaker in the light for about one week. When the cells have grown to sufficient concentration, you will test for nitrate reductase activity in the transformed mutant and in untransformed cells. Nitrate reductase is an inducible enzyme. It is synthesized only in the presence of nitrate and its synthesis is inhibited by ammonium in many systems. Therefore, since the mutant cells are grown in an ammonium medium, which would inhibit the synthesis of nitrate reductase, these cells will be transferred to a nitrate medium several hours before the enzyme assay is performed.

MATERIALS

PART I
Equipment
15-mL sterile polypropylene tubes
   with plug seal caps
pipettors and tips (sterile)
0.45-0.52 mm glass beads (300 mg)
   preweighed in test tubes (sterile)
vortex mixer
plastic petri dishes containing
   SGII-NO$_3$ medium + 1.5% agar
   in 9-cm plastic petri dishes
plastic petri dishes containing
   SGII-NH$_4$ medium + 1.5% agar
   in 9-cm plastic petri dishes
glass spreader
laminar-flow hood (or sterile area)
antibacterial soap
growth chamber, 25°C, 14 h. light, 10 h. dark, 30 W fluorescent bulbs

Solutions
SGII-NO$_3$ medium (see appendix)
20% polyethylene glycol (av. Mol. Wt. = 8000)
plasmid DNA (pMN24) containing the
   nitrate reductase gene (see Appendix G)
70% ethyl alcohol

Plant material
Chlamydomonas reinhardtii
(nit1-305 cw-15) grown on SGII-NH$_4$
medium and resuspended in SGII-NO$_3$
for media) medium (see Appendix G for recipes)

PART II
Equipment
test tubes and racks
pipettors and tips
clinical centrifuge
spectrophotometer and
   tubes or cuvettes

Solutions
Reaction buffer (pH 7.1): 250 mM Na$_2$PO$_4$,
   50 mM NaNO$_3$, 50 μM dithiothreitol (DTT)
1% sulfanilamide
0.2% NEED
25 mM KNO$_2$ (50 nmol/mL)
Nystatin (5000 units/mL), store on ice

Plant material: cultures of mutant and transformed cells in nitrate buffer

IMPORTANT: Although this is a minimum-risk experiment (Section III-C, Fed. Reg. May 87, 1986, p. 16961, requiring only the approval of the institution), the use of recombinant DNA requires that containment procedures be followed. THE FOLLOWING RULES MUST BE OBSERVED DURING CLASS:

1. Work surfaces must be decontaminated regularly and after any spill of viable material. If you should spill a sample containing plasmid DNA, notify your instructor.
2. Pipeting by mouth is prohibited. Mechanical pipeting devices must be used.
3. After handling materials containing recombinant DNA molecules, your hands should be carefully washed.
4. A laboratory coat should be worn when you are working with recombinant DNA materials.
5. All contaminated liquid or solid wastes must be decontaminated (by autoclaving) before disposal.

PART I. The Transformation of a Mutant *Chlamydomonas*. The gene for nitrate reductase, contained in a plasmid, will be used to transform the mutant *Chlamydomonas*, which lacks a functional nitrate reductase gene (Fig. 13-1).

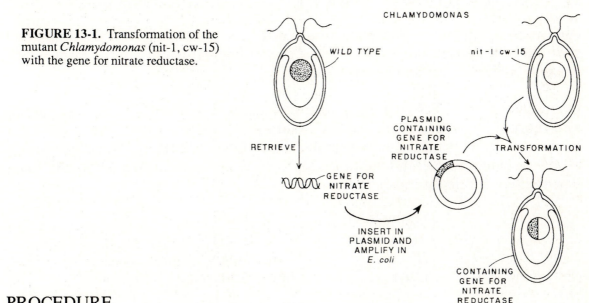

FIGURE 13-1. Transformation of the mutant *Chlamydomonas* (nit-1, cw-15) with the gene for nitrate reductase.

PROCEDURE

(1) Obtain three sterile 15-mL polypropylene tubes. Label one tube "+DNA" and two tubes "-DNA." Do not remove the caps except under sterile conditions as doing so would allow microorganisms to enter and contaminate the tubes.

(2)   Turn on the blower to cause a current of air to flow out of the sterile hood and turn off the ultraviolet (UV) lamps (if present). The blower acts to bring filtered air into the sterile area and the UV lamps prevent the growth of microorganisms. UV light is harmful; avoid looking directly at the UV light and never work at the sterile hood with the UV lamps turned on. Wear a lab coat while performing the following steps. Wash your hands with antibacterial soap.

(3)   Keep everything very clean. Use previously sterilized materials or sterilize all implements by dipping the part to come in contact with the cells in 70% ethyl alcohol for 10 seconds, then flaming off the alcohol with an alcohol lamp or Bunsen burner.

(4)   Working at the laminar-flow hood, add 0.3 mL of the *Chlamydomonas* mutant to be transformed to each of the tubes. The cells have been resuspended in medium containing nitrate (and no ammonium). Add 0.1 mL of 20% polyethylene glycol, which may make the plasma membrane more permeable and the transformation more efficient, and a premeasured amount (300 mg) of glass beads to each tube. Flame the mouth of the test tube containing the beads before pouring the beads into each tube.

(5)   Add 1 μg (your instructor will tell you what volume to add) of the DNA plasmid to the +DNA tube only and cap the tubes.

(6)   Agitate each tube on a vortex mixer at top speed for 15 seconds. It is important that the tubes be capped during this step since biological agents may be spread by aerosols, which are created during vortexing.

(7)   Obtain two petri dishes containing agar medium with nitrate (and no ammonium) and label them +DNA and –DNA. Obtain a third petri dish containing agar medium with NH_4 (and no nitrate) and label it –DNA. At the sterile hood, transfer **all** of the liquid from each tube to an appropriate petri dish, using a pipettor and a new sterile pipet tip for the three treatments:

Treatment	Medium
+DNA	SGII-NO_3
–DNA	SGII-NO_3
–DNA	SGII-NH_4

Dip a glass spreader in 70% ethyl alcohol, flame off the alcohol, then cool the glass spreader on the agar at the edge of the plate. Spread the cells around the surface of the agar with the spreader and immediately replace the lid of the petri dish.

(8)   Place your three labeled petri dishes in a growth chamber at about 25 ° C. The lab staff will seal your dishes with Parafilm on the next day.

(9)   Examine your petri dishes after one and two weeks. Transformed cells may be seen after one week with a dissecting microscope. After two weeks, look for the development of colonies, which will be seen as dark green spots on the agar. Count the colonies and record the number in the chart on page 121.

PART II. An Assay for Nitrate Reductase Activity.

Once the colonies have been established (after about two weeks), the instructor (or a student volunteer) will transfer a single colony of transformed cells to liquid SGII-NO$_3$ medium under sterile conditions. The culture will be allowed to grow in the light on a shaker for about one week (the instructor will make sure the cells have grown to the proper concentration). In addition, mutant cells (nit1-305), which are defective in the gene for nitrate reductase, will be grown on SGII-NH$_4$ medium. Both will be transferred to a nitrate containing buffer before class. You will test for nitrate reductase activity in both types of cells, by measuring the **nitrite** concentration (nitrite is the initial product of the reaction) of an assay medium before and after the reaction. The cells will first be permeabilized with Nystatin to allow the nitrite which is produced to leak into the medium. Nitrite reacts with sulfanilamide and N-1-naphthylethylenediamine dihydrochloride (NEED) to form an azo dye, which may be measured spectrophotometrically. The readings for the experimental treatments may be compared with those for a series of known nitrate concentrations, making up a standard curve.

PROCEDURE

(1)   Label six test tubes (13 x 100 mm) for the following treatments: Mutant cells, 0 and 10 minutes (Treatments 1 and 2); Transformed cells, 0 and 10 minutes (3 and 4); Boiled transformed cells, 0 and 10 minutes (5 and 6).

(2)   Add 2 mL of the appropriate cells to each test tube. Boil treatments 5 and 6 by placing the tubes in a beaker of boiling water for two minutes.

(3)   Add 20 μL of the Nystatin to each tube and shake thoroughly. The reaction time will begin after the cells have incubated in the Nystatin for one minute; after the minute is up, add 2 mL of the sulfanilamide solution to the 0 time controls (treatments 1, 3, and 5). Allow the remaining treatments to incubate at room temperature in the light for an additional 10 minutes. At the end of 10 minutes, add 2 mL of the sulfanilamide solution to treatments 2, 4, and 6.

(4)   Set up six test tubes (#7-12) for the standard curve according to the protocol in Table 13-1. Add 2 mL of the sulfanilamide to tubes 7 to 12.

TABLE 13-1. Protocol for nitrite standard curve (tubes 7-12), using a standard nitrate solution of 25nmol/mL.

Tube #	Nitrite (nmol)	Nitrite Solution (mL)	Water (mL)
7	0	0	2.0
8	10	0.4	1.6
9	20	0.8	1.2
10	30	1.2	0.8
11	40	1.6	0.4
12	50	2.0	0

(5)   Add 2 mL of the NEED solution to all tubes and wait five minutes to allow a pink color to develop. Sediment the reaction mixes (1-6) in a clinical centrifuge for five minutes to remove denatured cells.

(6)   Transfer the supernatant from each test tube into a labeled spectrophotometer tube or cuvette and read the absorbance (at 540 nm). Use nitrate buffer to set the zero when reading treatments 1 to 6. Use tube 7 as the blank to read the standard curve tubes (7-12). Record the data in the chart on page 121.

CALCULATIONS

Calculate the amount of nitrate reductase activity as nanomoles of nitrite produced (during your ten minute reaction time) per mL of cells. Use Eq. (13-1) to determine the nanomoles of nitrite produced (in 10 minutes) per milligram of chlorophyll. Use the chart on page 121.

$$\text{nmol NO}_2^-/\text{mg} = \frac{\text{nmol NO}_{2\,(10\text{ min})} - \text{nmol NO}_{2\,(0\text{ min})}}{2\text{ mL of cells}} \qquad \text{Eq. (13-1)}$$

TEXT REFERENCES

Galston, A. W., P. J. Davies, and R. L. Satter, *The Life of the Green Plant* (3rd ed.), pp. 177-184. Englewood Cliffs, NJ: Prentice Hall, 1980.

Salisbury, F. B. and C. W. Ross, *Plant Physiology* (4th ed.), pp. 295-300. Belmont, CA: Wadsworth Publishing Co., 1991.

Taiz, L. and E. Zeiger, *Plant Physiology*, pp. 302-4. Redwood City, CA: Benjamin/Cummings Publ. Co., Inc., 1991.

FURTHER READING

Kindle, K. L., "High-frequency nuclear transformation of *Chlamydomonas reinhardtii*," *Proc. Natl. Acad. Sci. USA* (1990), 87:1228-1232.

Kindle, K. L., R. A. Schnell, E. Fernández and P. A. Lefebvre, "Stable nuclear transformation of *Chlamydomonas* using the *Chlamydomonas* gene for nitrate reductase," *J. Cell Biol.* (1989), 109:2589-2601.

Rajasekhar, V. K. and R. Oelmüller, "Regulation of induction of nitrate reductase and nitrite reductase in higher plants," *Physiologia Plantarum* (1987), 71:517-521.

EXPERIMENT 13

Name _____

Date _____

PART I.

Note the number of colonies in each petri dish and record the data in the table below (it is not necessary to count the colonies present on the SGII-NH$_4$ medium; merely note whether growth was observed). Considering that the cell suspension you used contained about 5 x 10^7 cells (your instructor can give you a more **exact** figure), what was the efficiency of transformation (recorded as the number of transformations/cell)?

Treatment	Medium	Number of Colonies	Efficiency of Transformation
+ DNA	SGII-NO$_3$		
– DNA	SGII-NO$_3$		
– DNA	SGII-NH$_4$		

PART II.

Record the absorbance for each of the treatments in the charts below.

#	Treatment	Reaction Time	A_{540}	Nitrite (nmol)	Nitrite (nmol/mL) after 10 min
1	Mutant	0 min			
2	Mutant	10			
3	Transformant	0			
4	Transformant	10			
5	Boiled Transformant	0			
6	Boiled Transformant	10			

#	7	8	9	10	11	12
Nitrite (nmol)	0	10	20	30	40	50
A_{540}						

Plot the readings for tubes 7-12 on the graph on the next page.

Find the readings for tubes 1-6 on the graph and determine the nitrite concentration of each. Finally, determine the rate of nitrite production as nmol/mL in 10 minutes using Eq. (13-1).

Amount nitrate/tube (nmol)

QUESTIONS

1.   What is the fate of the nitrate ions which are taken up by the root cells of higher plants?

2.   Why did you do a -DNA treatment on the SGII-NO$_3$ medium? Considering that the nit1-305 mutant has a single point mutation, could you explain the growth of a colony of mutant *Chlamydomonas* on medium that contains nitrate but does not contain ammonium?

3.   Why did you include a treatment of untransformed cells (without any added DNA) grown on SGII-NH$_4$ medium?

4.   Compare the nitrate reductase activity in the transformed and untransformed cells. Why were the transformed cells able to produce nitrite?

5.   Explain why it was necessary to resuspend the untransformed mutant in a nitrate buffer before the enzyme assay was performed. How might you test for the requirement for nitrate for the induction of nitrate reductase in these cells?

6.   Why did you do an enzyme assay of boiled transformed cells?

EXPERIMENT 14

Ion Uptake by Potato Tuber Discs

INTRODUCTION

Plants must extract the ions needed for growth from their environment. Although roots are the usual absorbing organs, other tissues are also able to take up salts. You will use potato tuber tissue in an attempt to determine if phosphate ions are taken up by active or passive transport.

Passive transport is defined as the movement of ions across a membrane down an electrochemical gradient, while **active transport** is the movement of ions against an electrochemical gradient at the expense of metabolic energy (ATP). Active transport of ions usually occurs through carriers, which are proteins located in cell membranes. In the case of some anions, like phosphate, active transport is likely dependent on the H^+-ATPase located at the plasma membrane. One proposed model for phosphate uptake is illustrated in Fig. 14-1.

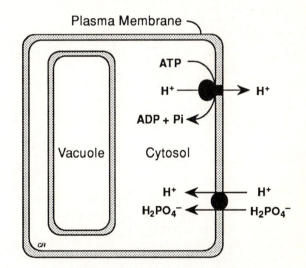

FIGURE 14-1. A proposed model for the transport of phosphate ions.

Potato tuber discs that have been aged (cut and incubated in aerated water for about 24 hours) are metabolically more active than freshly cut discs (see Expt 8). The increase in the rate of respiration (and therefore ATP production) provides us with the opportunity to examine rates of ion uptake in relation to available energy. The effects of respiratory inhibitors, uncouplers and low temperature on ion uptake are also useful indications of the nature of ion uptake in potato tuber tissue, but measurement of the membrane potential and internal concentration would be necessary to truly demonstrate active uptake.

Substantial quantities of most salts needed and accumulated by plants are already present in any tissue we might use for a study of ion accumulation. Chemical measurement of an increase in the amount of a particular ion over a short period of time would be difficult since a tiny amount taken up would not be measurable in the presence of the large amount already present. The use of radioactive isotopes provides an easy and accurate method of measuring ion accumulation. The radioactive atoms are the same as stable ones except for a slight difference in mass; the chemical properties in particular are identical. Very small amounts of the radioactive element of interest can be measured with a Geiger counter or by liquid scintillation counting.

In Geiger-Müller (G-M) counting, the radiation (in this case, a β particle) enters a chamber containing an ionizable gas. Electrons, resulting from collision of the β particle with the gas, are attracted to and affect the electrical properties of a positively charged electrode. The resulting electrical pulses are counted by the electronic system of the machine. The sample sits below the chamber containing the charged wire; only those particles moving in the direction of the chamber are counted. Data is recorded as counts per minute (CPM). Liquid scintillation counting is discussed in the Introduction of Expt. 6.

MATERIALS

Equipment
plastic trays lined with absorbent paper
latex gloves
lab coats
safety glasses
forceps
150-mL beakers
shaker
ice
No. 3 filter paper
automatic pipets
disposable pipet tips
planchets or scintillation vials
table-top planchet counter or
    liquid scintillation spectrometer
nitrile or toluene resistant gloves
    (for wear when using toluene containing
    scintillation cocktails)
Plexiglas shielding (optional)

Solutions
0.02 mM KH_2PO_4, pH 6 containing 0.2 mM $CaSO_4$
0.2 mM KH_2SO_4, pH 6 containing 0.2 mM $CaSO_4$
1 mM KH_2PO_4, pH 6 (rinse)
1 M KH_2PO_4, pH 6 (for cleanup)
1 mM CCCP in DMSO NOTE: toxic
60 mM KCN NOTE: toxic; use with care
$KH_2{}^{32}PO_4$, 1.8 μCi (6.66 x 10^4 Bq)/mL (shielded)
scintillation cocktail in Repipet (see Appendix G)

Plant material
potato (*Solanum tuberosum*) discs cut with a
No. 11 cork borer; "aged" discs have been
incubated in aerated water for 24 h.

Cleanup
liquid-radioactive waste containers (separate
    containers for KCN and CCCP)
solid-radioactive waste container

IMPORTANT: Great care must be taken when using radioactive isotopes to avoid transferring any isotopes to your hands, clothes, or equipment. The following precautions are mandatory when a high-energy β radiation emitter like ^{32}P is used. PLEASE OBSERVE THE FOLLOWING PRECAUTIONS FOR THE HANDLING OF RADIOACTIVE MATERIALS DURING CLASS!

1. Never pipet radioactive solutions by mouth. Automatic pipetting devices must be used to transfer radioactive solutions.
2. Wear latex gloves and a disposable lab coat when radioactive solutions are used.
3. Label all materials which come in contact with the radioactive solution with warning tape. Remove the tape when these items are decontaminated.
4. Carry out all work in a tray lined with and placed on absorbent paper so that minor spills will be contained and radioactivity will not be spread around the laboratory. Keep stock solutions in secondary containment and behind Plexiglas shielding.
5. Never hold radioactive material near one's eyes. The eye is especially sensitive to ionizing radiation.
6. Dispose of waste radioactive materials in the special containers provided, but do not put other, non-radioactive materials, in these containers.
7. Tell your instructor of any accidental spillage of radioactive solution!
8. Monitor all materials, equipment, and your bench during and after use. The instructor will monitor your hands and shoes before you leave class today to ensure that no radioactive materials are carried out of the classroom.
9. Read the appendix on radioactivity before beginning the experiment.
10. Always keep solutions containing ^{32}P behind Plexiglas shielding when quantities greater than $10 \mu Ci$ are used. The use of shielding is also recommended when smaller quantities of ^{32}P are used, but its use requires practice and care.

PROCEDURE

You will be issued the proper amount of radioactive phosphate when your bathing solution is prepared. A 0.1 mL sample of your bathing solution will be counted to determine the CPM that are equivalent to the amount of phosphate in your solution. The potato discs will be added to the bathing solution all at once, then removed at specific time intervals for counting. Use forceps when handling your radioactive discs; never touch them with your hands. Wear latex gloves, especially while transferring discs to the planchets. Each disc will be rinsed to remove the ^{32}P present in the cell walls (free space), then counted by G-M or liquid scintillation counting (ask your instructor which method you will use). The instructor will assign one of the following treatments to each group:

A. Fresh discs, 0.02 mM KH_2PO_4, room temperature.
B. Aged discs, 0.02 mM KH_2PO_4, room temperature for 30 minutes. Add 0.2 mL DMSO, then take the last three sets of samples. The added DMSO is a control for Treatment F. Be careful; DMSO is a strong solvent.

C.   Aged discs, 0.02 mM KH_2PO_4 at about 0 °C for 30 minutes (until after the fourth set of discs is removed). Warm the solution to room temperature and then take the last three sets of samples.

D.   Aged discs, 0.02 mM KH_2PO_4 at room temperature for 30 minutes. Cool to about 0 °C and then take the last three sets of samples.

E.   Aged discs, 0.02 mM KH_2PO_4, room temperature, for the first 4 measurements, then add 0.2 mL of 60 mM potassium cyanide (KCN) to the beaker and take the last three measurements. KCN is a respiratory inhibitor and will decrease the amount of ATP produced. It is also toxic; wear latex gloves, use an automatic pipettor, and be careful.

F.   Aged discs, 0.02 mM KH_2PO_4, room temperature, for the first 4 measurements. Then add 0.2 mL of 1 mM carbonyl cyanide m-chlorophenol hydrazone (CCCP) to the beaker and take the last three measurements. CCCP is an uncoupler that dissipates H^+ gradients; consequently, it will also reduce the formation of ATP coupled to respiration. CCCP is toxic; use an automatic pipettor and be careful. Wash your skin immediately if it comes in contact with this solution.

G.   Aged discs, **0.2** mM KH_2PO_4, room temperature. The phosphate concentration is ten times as great as in other treatments.

H.   Fresh discs, 0.02 mM KH_2PO_4, room temperature. Blot the discs but omit the rinses. This treatment will demonstrate the uptake into the free space.

(1)   Put 20 mL of the 0.02 mM KH_2PO_4 solution (use 0.2 mM for treatment G) in a 400-mL beaker. Also prepare three 150-mL plastic beakers for the rinsing treatments (step 8); each should contain about 100 mL of 1 mM KH_2PO_4 solution.

(2)   Label 16 planchets or scintillation vials (on the cap): two for each time interval (5, 10, 20, 30, 40, 50 and 60 minutes; see step 7), one for the 0.1 mL sample of solution (see step 4) and one for background (to account for incident radiation).

(3)   Put on latex gloves, safety glasses and a lab coat. Ask the instructor to add 1 mL of carrier-free ^{32}Pi solution containing 4 x 10^6 DPM (1.8 μCi or 6.66 x 10^4 Bq) to your beaker. Mix gently. Keep the beaker in a tray and transport the entire tray whenever you must move from one workspace to the next. If Plexiglas shielding is available, keep your tray and beaker behind the shielding whenever possible. The instructor will turn on the portable Geiger counter; monitor your hands and equipment frequently during the course of the experiment.

(4)   Use an automatic pipettor to transfer a 0.1-mL sample of the solution to the planchet or vial marked "0.1-mL sample" (put the used pipet tip in the "solid radioactive waste"). You will use the counts per minute (CPM) for this quantity of bathing solution in your calculations.

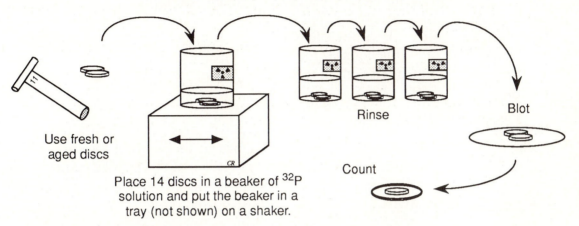

Remove discs at 5, 10, 20, 30, 40, 50, and 60 minutes.

Use fresh or
aged discs

Place 14 discs in a beaker of ^{32}P
solution and put the beaker in a
tray (not shown) on a shaker.

Rinse

Blot

Count

FIGURE 14-2. Procedure for measurement of phosphate uptake by potato tuber discs. Carry out all work in or over a tray lined with absorbent paper. The instructor will add the labeled phosphate. Remember to take a 0.1 mL sample of the bathing solution.

(5)   Select 14 discs from either the container marked "fresh discs" or "aged discs" and rinse the discs with distilled water. It is important that all the discs be of the same size and thickness; variations in surface area will greatly affect the rate of uptake.

(6)   Making sure to **record the time**, add all 14 discs to your solution and carefully place the beaker on a shaker set on low speed to gently stir the discs and solution.

(7)   When removing discs from your beaker, turn off the shaker and handle the discs with forceps only; do all further manipulations of the discs over your trays. Remove sets of two discs after the following time intervals: 5, 10, 20, 30, 40, 50 and 60 minutes (see Fig. 14-2).

(8)   As each set is removed, rinse the discs according to the following schedule:

Rinse #	Solution	Duration
1	1.0 mM KH$_2$PO$_4$	1 minute
2	1.0 mM KH$_2$PO$_4$	2 minutes
3	1.0 mM KH$_2$PO$_4$	1 minute

Using forceps, remove each disc from the last rinse and blot it on No. 3 filter paper. Transfer each disc to the appropriately labeled planchet or vial. One student should carry out the rinsing and blotting procedure to ensure uniformity in results.

(9)   If a table top G-M counter is available, count each disc separately, either during the experiment or after all the discs have been removed from solution. Also count your background planchet and the one containing 0.1 mL of solution. Be careful when

using the counter. The window of the G-M tube is a delicate membrane which is easily broken. Also, try not to spill radioactive fluids or drop radioactive discs on the planchet holder or other parts of the counter. If you do, **please** tell the instructor. Use the same counter for all your measurements and save the discs until all the counting has been completed. Record your data in the chart provided on page 131.

If you are using liquid scintillation counting, add 5 mL of the scintillation cocktail to each vial and replace the cap. Wear nitrile gloves if the scintillation cocktail you are using contains toluene. The instructor will collect the vials and begin the scintillation counting. The data will be available when the counting is completed; ask your instructor where and when to find the posted data.

CLEANUP

Pour the used radioactive solution into the "liquid radioactive waste." Solutions containing KCN or CCCP should be placed in separate labeled waste containers. Rinse the beakers with **1 M** KH_2PO_4 and add this rinse to the radioactive waste container. Now rinse each beaker 10 times in tap water, prior to the usual wash with detergent. All potato discs, the used planchets and the No. 3 filter paper should be placed in the "solid radioactive waste." Wash all other materials carefully. The instructor will monitor all materials for radioactivity at the end of the class.

CALCULATIONS

Knowing the counts per minute (CPM) for each treatment and the concentration of phosphate in the bathing solution, you should be able to calculate the nanomoles of phosphate taken up by each potato disc. The amount of phosphate added as $KH_2{}^{32}PO_4$ was negligible. The concentration of phosphate in the bathing solution was 0.02 mM (except for treatment G), which is equivalent to 0.02 μmol/mL and to 0.002 μmol or 2 nmol/0.1 mL.

Now if, **for example,** you determined that 0.1 mL of bathing solution contained 1000 CPM (after the correction has been made for background incident radiation), then 1000 CPM would be equivalent to 2 nmol of phosphate. (Note that the calculation is different for the 0.2 mM phosphate solution used in treatment G.) If the average observed CPM for the discs removed at 5 minutes were 250 CPM (after the background counts were subtracted), then **for this example:**

$$250 \text{ CPM} \times \frac{2 \text{ nmol of phosphate}}{1000 \text{ CPM}} = 0.5 \text{ nmol phosphate per disc in 5 minutes.}$$

Calculate the amount of phosphate taken up for each time period and plot the activity incorporated (nmol/disc) into the tissue against time on the graph on page 132. Determine the rate of uptake as nmol/disc-h.

For treatment H, you want to determine the **amount** taken up into free space rather than the rate of uptake. Since you did not rinse the discs, they should all contain some phosphate in the cell walls and intercellular spaces (free space). Do the calculations as described above and plot the nanomoles of phosphate against time. Construct the best fit line and extrapolate the line back to the y-axis. The value at 0 time will be equivalent to the amount of phosphate present in the free space of the discs.

TEXT REFERENCES

Galston, A. W., P. J. Davies, and R. L. Satter, *The Life of the Green Plant* (3rd ed.), pp. 184-195. Englewood Cliffs, NJ: Prentice Hall, 1980.

Salisbury, F. B. and C. W. Ross, *Plant Physiology* (4th ed.), pp. 145-159. Belmont, CA: Wadsworth Publishing Co., 1991.

Taiz, L. and E. Zeiger, *Plant Physiology*, pp. 120-133. Redwood City, CA: Benjamin/Cummings Publishing Co., Inc., 1991.

FURTHER READING

Dunlop, J. "Phosphate and membrane electropotentials in *Trifolium repens* L.," *Journal of Experimental Botany* (1989), 40:803-807.

Ulrich, C. I. and A. J. Novacky, "Extra- and intracellular pH and membrane potential changes induced by K^+, Cl^-, $H_2PO_4^-$, and NO_3^- uptake and fusicoccin in root hairs of *Limnobium stoloniferum*," *Plant Physiol.* (1990), 94:1561-1567.

EXPERIMENT 14

Name _____

Date _____

RESULTS

Record your data in the chart below, subtract the background counts and determine the nmol/disc of phosphate taken up for each treatment.

Treatment _____

Treatment	CPM		Average CPM	−Background = True CPM	Pi (nmol/disc)
Background					
0.1 mL					
5 min					
10 min					
20 min					
30 min					
40 min					
50 min					
60 min					

Draw the curve (not necessarily through the origin) for your treatment on the graph on the next page. From the slope of the best straight line on the graph you construct, express the rate of uptake in units of nmol/disc-h. If you changed treatments after 30 minutes, you must determine two rates of uptake.

Rate(s) of uptake _____ / _____

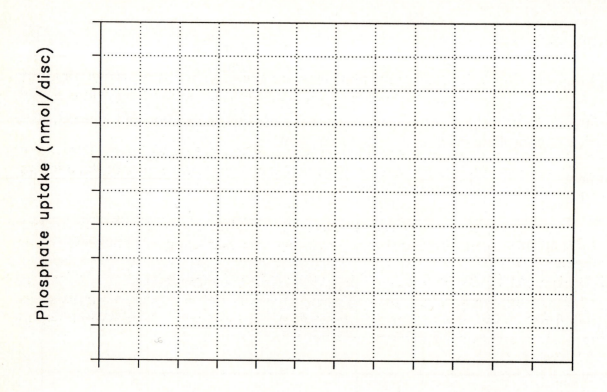

Time (minutes)

Gather results for the rates of uptake for the other treatments from the rest of the class and include them here.

#	Treatment	Rate (nmol/disc-h)	#	Treatment	Rate (nmol/disc-h)
A	Fresh		E	Aged /+KCN	/
B	Aged /+DMSO	/	F	Aged /+CCCP	/
C	Aged, 0°C /Aged	/	G	10X Concentration	
D	Aged /Aged, 0°C	/	H	Free Space (nmol)	

QUESTIONS

1. Does the class data imply that phosphate uptake is an active or passive process? Describe how each treatment enabled you to answer this question. Be complete and cite the data.

2. Referring to the model for phosphate uptake given in Fig. 14-1, explain how the addition of the uncoupler CCCP could affect the active uptake of phosphate in two ways.

3. Are there any other explanations for the results observed? What further information would be required to demonstrate that uptake is either active or passive?

4. The **amount** of phosphate present in the free space of unrinsed potato discs was determined using the data from treatment H. How does this information help to explain irregularities in your graphs for treatments A through G?

EXPERIMENT 15

Ion Transport and Electrochemical Potentials

INTRODUCTION

By now you should be aware that ions cross membranes in two ways: actively and passively. The active mechanism (sometimes involving an ion pump) uses energy to move ions across a membrane in a direction opposite to the direction they would go (i.e., against an electrochemical gradient) if they were allowed to diffuse freely. Ion pumps usually consist of a "carrier protein" with which the ion combines to cross the membrane.

In the case of passive ion movements, the membrane potential (the electrical potential difference across a membrane) can have a great effect on ion movement. Since ions cross the membrane as charged particles, their tendency to move will be influenced by the electric field across the membrane. Thus an ion diffuses passively both down a concentration gradient (as does a neutral molecule) **and** down an electrical potential gradient. The sum of these two gradients is known as the **electrochemical potential gradient**; it determines the direction in which passive diffusion takes place. If the net movement of an ion is in the opposite direction, then active transport must be involved (i.e., there is active as well as passive transport).

Detopped roots, if supplied with salts in a well-aerated bathing solution, will continue to exude solution from the xylem for long periods. This phenomenon is known as root exudation, and the driving force is called root pressure. Active uptake of salts into the xylem of the root leads to a more negative water potential in the xylem, causing water to move from the bathing solution into the stele. It is clear that there must be an active inward movement of some salts because the salt concentration in the exudate is generally greater than that in the bathing solution. To determine if a particular ion is being actively transported, it is necessary to measure the **electrochemical potential gradient** between exudate and bathing solution for that ion. You will determine if Cl^- is actively transported by measuring:

    a)   the concentration of the Cl^- ion in both the exudate and the bathing solution in Part I.

    b)   the electrical potential difference between the exudate and the bathing solution in Part II.

Then, you will use Eq. (15-1) to calculate the electrochemical potential difference. With this information you can determine whether the transport of the Cl^- ion into the xylem of a *Xanthium* or bean root is active or passive.

MATERIALS

Equipment	Solutions
Tygon tubing	3 M KCl
aeration system	100 mM KCl
beakers	10 mM KCl
millivoltmeter (pH meter) with	1 mM KCl
calomel reference electrode and	0.1 mM KCl
Ag/AgCl wire electrode (see	bathing solution:
Appendix G)	0.5 mM NaCl and
salt bridges stored in 3 M KCl	0.5 mM KCl
3-cycle log paper for standard curve	

Plant Material
Cocklebur (*Xanthium strumarium*) or bean (*Phaseolus vulgaris* var. Redcloud) plants, 5 weeks old, detopped and kept in aerated bathing solution for 24 to 48 h.

PART I. The Determination of Cl^- Concentration.

Detopped *Xanthium* or bean roots, which have been kept in an aerated salt solution overnight, will be provided. A piece of tubing has been attached to the stump to collect the exudate.

A standard curve will be used to determine the Cl^- concentration. The electrical potential between a reference electrode and a silver/silver chloride (Ag/AgCl) electrode is related to the log of the Cl^- concentration of the solution in which the Ag/AgCl electrode is immersed. The potential difference has been measured for a range of standard Cl^- solutions using the methods described below. The log of the Cl^- concentration (in mM) of each standard solution has been plotted against the measured potential difference (in mV) on three cycle semi-log paper.

PROCEDURE

(1)   The chloride concentration in the bathing solution is known to be about 1.0 mM Cl⁻.
       Check this concentration by measuring the potential difference between a reference
       electrode and a silver/silver chloride (Ag/AgCl) wire when both are placed in
       contact with the bathing solution. The Ag/AgCl wire may be placed directly into
       the bathing solution and the reference electrode may be put in contact with the
       solution using a salt bridge (Fig. 15-1).

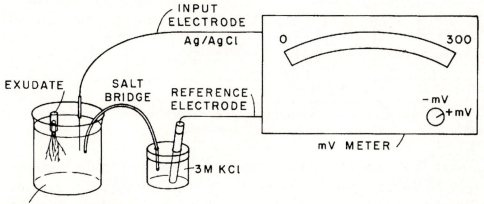

FIGURE 15-1. Circuit for concentration determinations. The electrodes should first be
connected to the bathing solution (as shown) for step (1), then both electrodes should be
placed in the exudate for step (2). The millivoltmeter shown is a pH meter and must be
read on the expanded scale. If you use a different instrument, follow the special
instructions provided by your instructor.

Turn on the millivoltmeter and read the potential difference on the millivolt scale
(0-300 mV). Determine the Cl⁻ concentration of your bathing solution by finding
the correct point on the standard curve.

(2)   Find the Cl⁻ concentration of the exudate by the method described above. This time
       put both the Ag/AgCl wire and the salt bridge (connected to the reference
       electrode) into the exudate solution. Find the potential difference in mV and
       determine the Cl⁻ concentration of your exudate from the standard curve. Record
       your results in the chart on page 141.

PART II. The Determination of the Electrical Potential Difference. (The millivoltmeter
must be rezeroed when the electrode is changed; see Appendix G for instructions.)

PROCEDURE

(1)   Measure the electrical potential difference using two calomel electrodes and salt
       bridges. The salt bridge connected to the **reference** calomel electrode should be

placed in the bathing solution, while the salt bridge connected to the **input** calomel electrode should be placed in the exudate (Fig. 15-2).

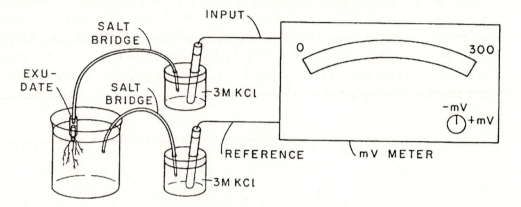

FIGURE 15-2. Circuit for determination of the electropotential difference between the xylem sap (exudate) and the bathing solution.

(2)   The electropotential difference between the exudate and the bathing solution can be read on the + or   millivolt scale of the meter. Record your reading in the chart on page 141.

CALCULATIONS

Calculate the net passive driving force on the ion, **the difference in electrochemical potential ($\Delta\mu$)**, from the following equation:

$$\Delta\mu = RT \ln C_{in}/C_{out} \; + \; z \, F\Delta E \qquad\qquad \text{Eq. (15-1)}$$

where:

R = the gas constant = 8.31 J/K-mol
T = temperature (K)
C_{in} = the concentration of Cl^- in the exudate fluid
C_{out} = the concentration of Cl^- in the bathing solution
z = the electrical charge on the ion = -1 for the chloride ion
F = Faraday's constant = 9.65×10^4 J/V-mol
ΔE = the electropotential difference = $E_{in} - E_{out}$ (in V)

Substitute your measured values (be sure to convert mV to V for ΔE) and the value for z into the above equation to determine the electrochemical potential difference ($\Delta\mu$ in J/mol) between the bathing solution and the exudate fluid for the Cl^- ion.

To establish that an ion moves across the cell membrane by active transport, it is necessary to show that the ion moves against its electrochemical potential gradient.

If the determined $\Delta\mu$ is equal to zero, then the electrochemical potential is the same in the exudate and the bathing solution. After the detopped roots were placed in the bathing solution, the Cl⁻ moved passively until equilibrium was reached.

A positive value means that the electrochemical potential of the ion in the exudate is greater than that in the bathing solution. This situation would give rise to a net passive efflux; the Cl⁻ ions would move from the exudate into the bathing solution. Therefore, to explain a greater concentration of Cl⁻ inside (a net influx of ion) under these conditions, it is necessary to postulate an **active** influx.

If the $\Delta\mu$ is negative, the net passive influx would be inwards. It would not be necessary to postulate an active uptake mechanism to explain a greater concentration of Cl⁻ inside.

TEXT REFERENCES

Galston, A. W., P. J. Davies, and R. L. Satter, *The Life of the Green Plant* (3rd ed.), pp. 185-191. Englewood Cliffs, NJ: Prentice Hall, 1980.

Salisbury, F. B. and C. W. Ross, *Plant Physiology* (4th ed.), pp. 145-159. Belmont, CA: Wadsworth Publishing Co., 1991.

Taiz, L. and E. Zeiger, *Plant Physiology*, pp. 81-86, 120-133, 138-143. Redwood City, CA: Benjamin/Cummings Publishing Co., Inc., 1991.

FURTHER READING

Hedrich, R. and J. I. Schroeder, "The physiology of ion channels and electrogenic pumps in higher plants," *Annu. Rev. Plant Phys. Plant Mol. Bio.* (1989), 40:539-569.

Nicholls, D. G., *Bioenergetics*. New York: Academic Press, 1982.

Wood, P. M., "What is the Nernst equation?" *Trends Biochem. Sci.* (1985), 10:106-108.

EXPERIMENT 15   Name _____
                                        Date _____
RESULTS

Record your readings here. Calculate the electrochemical potential difference for your plant. Show your calculations in the space below.

Solution	mV Reading	[Cl⁻]	Electrical Potential Difference (ΔE) in V	
Bathing Solution			Electrochemical Potential Difference ($\Delta \mu$) as J/mol	
Guttation Fluid				

QUESTIONS

1.   Is it necessary to postulate an active transport mechanism between the bathing solution and the xylem? Explain why or why not.

2.   Where in the root would you expect such a mechanism to be located?

EXPERIMENT 16

Translocation of Labeled Sucrose

INTRODUCTION

Translocation is the process by which photosynthetic carbohydrates from mature leaves (sources) are moved through the phloem to growing tips, roots, flowers and fruits (sinks). According to the **Münch pressure flow** model, these solutes (primarily sucrose) move through the phloem sieve tubes along a pressure gradient by mass flow. This pressure gradient is generated by the active uptake of sugars at the source by H^+-sucrose co-transport, subsequent water movement into the sieve cells by osmosis and the removal of sucrose (probably through plasmodesmata) and water from the sieve tube at the sink. If sucrose labeled with the radioactive isotope carbon-14 (^{14}C) is applied to the mature leaves of whole plants, the ^{14}C-sucrose will be translocated through the plant.

The pathway for sucrose movement from source to sink may then be observed by autoradiography. In autoradiography, labeled plants are gently pressed against X-ray film for a period of several weeks. During this time the ions produced by radiation will chemically alter the X-ray emulsion. The result will be dark areas on the developed film in the places where ^{14}C is present in the plant.

To test the model for the mass flow of sucrose and gain additional information about the mechanisms for phloem transport and loading, several experimental treatments are suggested. For example, the uncoupler 2,4-dinitrophenol (DNP), which dissipates H^+ gradients and inhibits the production of ATP, may be applied to the source leaf. DNP may also be applied along the translocation pathway to test for an energy requirement for translocation through the phloem. By heating the petiole, you may determine if the phloem cells are required to be living for translocation to occur.

143

MATERIALS

Equipment
Class 1:
latex gloves
disposable lab coats
plastic trays
plastic backed absorbent paper
lanolin in a 10 mL syringe
cover slips, No. 2
Bunsen burner and tongs
125-mL Erlenmeyer flasks
cotton
small paint brushes
carborundum

The next day:
latex gloves
10" x 12" sheets of #110 Index white paper
container for solid radioactive waste

Class 2:
1/2" foam pads, 10" x 12"
10" x 12" heavy cardboards
10" x 12" Kodak X-Omat RP film
aluminum foil
dark room with red safe light

Solutions
5 mM 2,4-dinitrophenol (DNP). NOTE: DNP is
    a toxic chemical; be careful
^{14}C-sucrose, 10μCi/mL
20 μM sucrose

Plant material
3-week old bean (*Phaseolus vulgaris*,
    var. Redcloud) plants, one kept
    in the dark for 24 h.
6-week old bean (*Phaseolus vulgaris*,
    var. Redcloud) plants with
    developing fruit

masking tape (cut into small pieces)
plant press, blotters and spacers
Drying oven at 80°C

After 4 weeks:
dark room with red safe light
X-ray developer and fixer

IMPORTANT: Great care should be taken when using radioisotopes to avoid transferring any isotopes to your hands, clothes or equipment. The following precautions are mandatory when using a low level β-radiation emitter like ^{14}C. PLEASE OBSERVE THESE PRECAUTIONS FOR THE HANDLING OF RADIOACTIVE MATERIALS DURING CLASS!

1.   Never pipet radioactive solutions by mouth. Automatic pipetting devices must be used to transfer radioactive solutions.
2.   Carry out all work in a tray placed on absorbent paper so that minor spills will be contained and radioactive materials will not be spread around the laboratory. Also keep stock solutions in secondary containment.
3.   Label all materials which come in contact with the radioactive solution with warning tape. Remove the tape when these items are clean.
4.   Wear latex gloves and a lab coat when using radioactive solutions.
5.   Dispose of all radioactive waste materials in the special container provided, but do not throw other, non-radioactive, materials in this container.
6.   Tell your instructor of any accidental spillage of radioactive materials.

7.   Be sure to read Appendix E on radioactivity before beginning this experiment.
8.   The instructor will conduct wipe tests (see Appendix E) at the end of class to ensure that no area of the laboratory is contaminated.

PROCEDURE

The instructor will assign one of the following treatments to each group. Read all of the instructions before continuing.

A.   Two controls. Apply ^{14}C-labeled sucrose to one plant (A1) as described below and unlabeled sucrose to the second (A2).

B.   Leaf painted with 5 mM DNP (an uncoupler). After the leaf has been abraded (see below), wear latex gloves (DNP is a toxic chemical) and use a paint brush to apply the DNP to a primary leaf, then add the ^{14}C sucrose.

C.   Petiole painted with 5 mM DNP (an uncoupler). Abrade the petiole with carborundum (see below) and, wearing latex gloves, use a paint brush to apply DNP to the petiole of the primary leaf that will receive the labeled sucrose.

D.   Petiole heated to kill phloem. Heat the tip of a pair of tongs in the flame of a Bunsen burner, then lightly touch the petiole of the appropriate primary leaf with the hot tongs. The petiole will bend at the contact point, but should not be blackened or broken. The leaf will remain turgid during the experiment since the non-living xylem elements are not affected by the heat treatment.

E.   Plant previously kept in the dark for 24 hours to deplete the source. Keep the ^{14}C-treated primary leaf covered with foil during the next 24 hours.

F.   Apply the ^{14}C sucrose to a young leaf.

G.   Use two older bean plants with developing fruits. In one case apply the ^{14}C sucrose to a primary leaf (G1); apply the labeled sucrose to a younger, but fully developed leaf on the second plant (G2).

(1)   Remove a bean plant from its pot and gently wash the soil from the roots. Place the plant in a 125-mL Erlenmeyer flask containing about 100 mL of tap water, placing some cotton at the mouth of the flask to help keep the plant upright. Label the flask with your name and treatment.

(2)   Some plants will require a pretreatment. Follow the instructions above for your treatment, then abrade the surface of a small area of a primary leaf (for treatment F choose a newly formed leaf) using a brush and some carborundum. Then, in the abraded area, form a circular well with lanolin. The well should be deep enough to contain 0.1 mL of the applied sucrose solution and should be sealed sufficiently to prevent leaking. If necessary, use an applicator stick to seal the lanolin to the leaf.

(3)   Put on latex gloves before continuing. Place your plant in the tray lined with plastic backed absorbent paper. Ask the instructor to pipet 0.1 mL (containing 1 μCi) of

the ^{14}C-labeled sucrose into the well on your leaf. Seal the well with a cover slip to keep the sucrose from spilling out.

(4)   Return to the lab after one day and locate your plant. Wearing latex gloves and working over plastic-backed absorbent paper, use a razor blade to remove the portion of the leaf containing the lanolin well and radiolabeled sucrose. Place the removed portion of the leaf in the solid radioactive waste container.

(5)   Carefully tape the plant onto one 10" x 12" piece of white paper using **small** pieces of tape and making sure that most of the plant is visible. The radiation from ^{14}C is not sufficiently energetic to pass through the applied tape. Remember that you are particularly interested in observing likely sinks. Label the paper with your name and treatment. Place it between two blotters, then between two spaces in a plant press. The instructor will press the plants and dry them in an 80 °C oven overnight.

(6)   During the next class, place your dried plant (still taped to paper) in a one gallon clear plastic bag to prevent small pieces of dried plant material from escaping. In the darkroom, place a piece of X-ray film inside the bag, next to the dried plant. Then, form a sandwich with the following materials: one sheet of foam, the pressed plant and X-ray film, and one piece of heavy cardboard (Fig. 16-1). The foam backing allows for some "give," enabling the plant to be in intimate contact with the X-ray film without damaging the delicate surface of the film. Wrap these materials in foil and place your package in a plant press. The plants will be gently pressed and left in the dark for four weeks. During this time the ions produced by radiation will chemically alter the X-ray emulsion. The result will be dark areas on the developed film in the areas where ^{14}C sucrose is present.

(7)   At the end of four weeks, the instructor will return your developed X-ray film and dried plant (sealed in the plastic bag) to you. Examine the plants and the film for all of the treatments and note where in the plant the labeled sucrose has moved in each case. Use the chart on page 149 to record your observations.

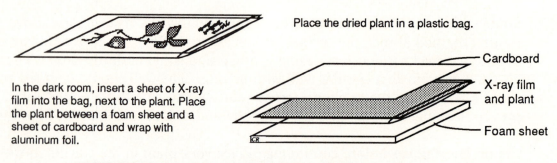

FIGURE 16-1. Procedure for exposing dried plant to X-ray film.

TEXT REFERENCES

Galston, A. W., P. J. Davies, and R. L. Satter, *The Life of the Green Plant* (3rd ed.), Chap. 8. Englewood Cliffs, NJ: Prentice Hall, 1980.

Salisbury, F. B. and C. W. Ross, *Plant Physiology* (4th ed.), Chap. 8. Belmont, CA: Wadsworth Publishing Co., 1991.

Taiz, L. and E. Zeiger, *Plant Physiology*, Chap. 7. Redwood City, CA: Benjamin/Cummings Publ. Co., Inc., 1991.

FURTHER READING

Crafts, A. S. and C. E. Crisp, *Phloem Transport in Plants*. San Francisco: W.H. Freeman and Co., 1971.

Cronshaw, J., W. J. Lucas and R. T. Giaquinta, eds., *Phloem Transport*. New York: Alan R. Liss, 1986.

Giaquinta, R. T., "Phloem loading of sucrose," *Annu. Rev. of Plant Physiol.* (1983), 34:347-387.

Richardson, M., *Translocation in Plants*. New York: St. Martin's Press, 1968.

Turgeon, R. "The sink-source transition in leaves," *Annu. Rev. Plant Physiol. Plant Mol. Biol.* (1989), 40:119-138.

EXPERIMENT 16

Name _____

Date _____

RESULTS

Record your observations in the table below.

#	Treatment	Observations of Autoradiographs
A1	Control	
A2	Control with Unlabeled Sucrose	
B	Leaf Painted with DNP	
C	Petiole Painted with DNP	
D	Petiole Heated to Kill Phloem	
E	Source Leaf Kept in Dark	
F	Labeled Sucrose on a Young Leaf	
G1	Plant with Fruit; Label on 1° Leaf	
G2	Plant with Fruit; Label on Younger Leaf	

QUESTIONS

1.   What governs the direction of flow in the phloem? Give specific examples from
      your observations.

2.   Can you say that energy (in the form of ATP) is required for any of the steps of
      translocation of sucrose through the plant? Explain.

3.   What two effects might the uncoupler have when applied at the source?

4.   What does treatment D tell us about the conditions needed for translocation in the
      phloem?

EXPERIMENT 17

Plant Tissue Culture

INTRODUCTION

If parenchyma cells are taken from a plant and placed in nutrient medium under appropriate sterile conditions, they divide and form a callus, which is a mass of undifferentiated cells. The tissue culture medium usually contains inorganic elements (as salts), with greater amounts of the major nutrients than micro-nutrients, and various organic compounds. As the callus does not photosynthesize, a carbohydrate such as sucrose must be provided. Vitamins (especially thiamine) and plant hormones (auxin and a cytokinin) are also included, as callus is unable to synthesize these compounds.

The plant hormone auxin is known to control cell enlargement and promote cell division in tissue culture. Kinetin, one of the group of hormones called cytokinins, induces cell division in conjunction with auxin. Both are needed for a healthy plant callus to develop. Roots and shoots can differentiate from the unorganized callus; this **development** of the callus is controlled by the hormonal composition of the nutrient medium. The relative concentrations of these two growth substances determine how the callus will differentiate. A high ratio of auxin to cytokinin promotes root development while a low ratio of auxin to cytokinin promotes shoot development.

You will grow tobacco parenchyma cells and observe the development of callus and the differentiation of roots or shoots on three media which contain different ratios of the two plant hormones discussed above; the auxin content will be the same in all three while the cytokinin concentration will vary. This experiment is an example of the classic approach to tissue culture and is based on the original work of Murashige and Skoog.

MATERIALS

Equipment
sterile work area
No. 1 cork borer
dissecting needle
single edge razor blade or knife
antibacterial soap
spirit lamp or Bunsen burner
growth chamber at 28° C under
    light ($25\,\mu\text{E m}^{-2}\text{ s}^{-1}$, 10-h. photoperiod)

Solutions
15% chlorine bleach
70% ethyl alcohol
media (see Table 17-1) in sterile containers

Plant material
Tobacco (*Nicotiana tabacum*) stems

PROCEDURE

You will be provided with containers of autoclaved medium. In addition to the nutrients, 0.5% agar is added to solidify the medium, enabling the tissue to develop on top of the medium with an adequate supply of oxygen. The nutrients present in the media are listed in Table 17-1.

TABLE 17-1. The components of a nutrient media for tissue culture, adapted from Skoog and Miller (1965).

Major Salts	mg/L	Micronutrients	mg/L	Organics	mg/L
$CaCl_2$	333	H_3BO_3	6.2	*myo*-Inositol	100
KH_2PO_4	170	$CoCl_2 \cdot 6\,H_2O$	0.025	Nicotinic Acid	0.5
KNO_3	1900	$CuSO_4 \cdot 5\,H_2O$	0.025	Pyridoxine HCl	0.5
$MgSO_4$	181	KI	0.83	Auxin (IAA)	2
NH_4NO_3	1650	$MnSO_4 \cdot H_2O$	16.9	Kinetin	0–0.5
$FeSO_4 \cdot 7\,H_2O$	27.8	$Na_2MoO_4 \cdot 2\,H_2O$	0.25	Casein Hydrolysate	1000
$FeNa_2EDTA$	36.7	$ZnSO_4 \cdot 7\,H_2O$	8.6	Thiamine HCl	0.4
				Glycine	2
Adjust pH to 5.7-5.8				Sucrose	30 g/L

The salts were mixed in a dilute solution, then the agar was added. The agar was dissolved by autoclaving, then the solutions were put in labeled vials and autoclaved for 15 minutes. Three types of nutrient media are provided, varying in auxin and cytokinin content as listed in Table 17-2.

TABLE 17-2. Three nutrient media containing auxin and kinetin.

Label	IAA Content	Kinetin Content
Green	2 mg/L	0.02 mg/L
White	2 mg/L	0.2
Yellow	2 mg/L	0.5

(1)   Take one container of each type and label it with your name. Do not remove the cap except under sterile conditions, as doing so would allow microorganisms to enter and contaminate your culture.

(2)   If a laminar-flow hood is available, turn on the blower to cause a current of air to flow out of the sterile hood and turn off the ultraviolet (UV) lamps (if present). The blower acts to bring filtered air into the sterile area and the UV lamps prevent the growth of microorganisms. UV light is harmful; avoid looking directly at the UV light and never work at the sterile hood with UV lamps turned on. Wash your hands with antibacterial soap.

(3)   Wipe the surface of the work area with 70% ethyl alcohol. Keep everything very clean. Sterilize all implements by dipping the part to come in contact with the tissue in ethyl alcohol for 10 seconds, then allowing the ethyl alcohol to evaporate. If a spirit lamp or Bunsen burner is available, you may flame off the ethyl alcohol, but **be careful**. Be sure to keep the flame away from the container of alcohol.

(4)   Cut a 15-mm segment of tobacco stem internode from near the apex and place it in a 15% chlorine bleach solution for three minutes to sterilize the surface.

(5)   With a sterile cork borer, remove a cylinder from the center of the tobacco stem, making sure to select pith tissue only (the vascular tissue contains hormones and its presence could affect your results). Push the tobacco pith onto a sterile work area. Using a sterile knife, cut off and discard the ends, then cut the remaining cylinder into three equal slices (Fig. 17-1). Each slice should be about 2 to 3 mm thick. Always handle the tissue with sterile forceps, **never** with your fingers.

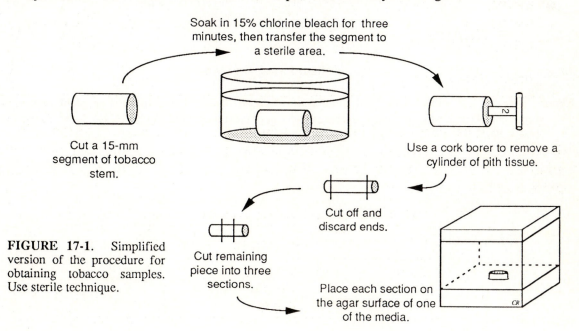

Soak in 15% chlorine bleach for three minutes, then transfer the segment to a sterile area.

Cut a 15-mm segment of tobacco stem.

Use a cork borer to remove a cylinder of pith tissue.

Cut off and discard ends.

FIGURE 17-1. Simplified version of the procedure for obtaining tobacco samples. Use sterile technique.

Cut remaining piece into three sections.

Place each section on the agar surface of one of the media.

(6)   Open a container of medium. The mouth of glass containers (but not plastic!) must be passed through the flame of a spirit lamp or Bunsen burner. Place one piece of tobacco pith on the agar with a sterile needle or forceps and press it into good contact with the agar. The tobacco disc should not be submerged in the agar, but set firmly onto the surface. Reflame the mouth of a glass vial; replace the lid. Repeat this procedure for each of the four types of media. Wipe the surface of the work area with 70% ethyl alcohol when you are finished. If UV lamps are present, turn them on.

(7)   Label the containers with your name and the date, then place them in an incubator at 28 ° C under dim light. Observe the development of the callus every two weeks for the next eight weeks.

During the developmental period you should observe the growth of the explants on a bi-weekly basis. Changes in color, shape and size, as well as the development of shoots or roots should be noted. Record your observations in the chart on page 155. If any microorganisms appear in your culture, it is because you were not sufficiently careful with your sterile technique. Should you need to redo your cultures, make sure that all tools are sterilized and that nothing gets into the sterile area to contaminate your material.

TEXT REFERENCES

Galston, A. W., P. J. Davies, and R. L. Satter, *The Life of the Green Plant* (3rd ed.), pp. 248-250 and 372-380. Englewood Cliffs, NJ: Prentice Hall, 1980.
Salisbury, F. T. and C. W. Ross, *Plant Physiology* (4th ed.), pp. 382-386. Belmont, CA: Wadsworth Publishing Co, 1991.

FURTHER READING

Murashige, T. and F. Skoog, "A revised medium for rapid growth and bioassays with tobacco tissue cultures," *Physiologia Plantarum* (1962), 15:473-497.
Skoog, F. and C. O. Miller, "Chemical Regulation of growth and organ formation in plant tissues cultured *in vitro*," in *Molecular and Cellular Aspects of Development*, ed. E. Bell, pp. 481-494. New York: Harper and Row, 1965.

EXPERIMENT 17

Name _____
Date _____

RESULTS

Record your observations in the chart below.

Starting date _____

Date	Green	White	Yellow

QUESTIONS

1. What is the effect of the various hormone levels? How does increasing the concentration of kinetin in the medium affect the differentiation of the tobacco callus?

2. What can tissue culture tell us about the genetic complement of individual plant cells? How do hormones control differentiation in plants?

3. From your reading, what practical uses are being made of plant tissue culture techniques?

EXPERIMENT 18

Plant Movements and the Differential Growth of Plants

INTRODUCTION

Plants are regarded as something "less alive" than animals by many people. This notion probably derives from the fact that plants do not appear to show striking responses to environmental stimuli. Hence the rapid movements of plants such as *Mimosa pudica* are regarded by many as extraordinary for a plant. Time lapse photography gives us quite a different picture. Nutation, phototropism, gravitropism, hydrotropism, nyctinasty and thigmonasty all cause a spectacular array of behavior. These differential growth responses and movements benefit plants in a number of ways (the ecological explanations will not be a concern here) and provide the physiologist with excellent systems for investigation of the physiological mechanisms involved.

Because data can be obtained rapidly and observations of plant growth and movements are easily made, such responses often serve as tools for the study of more fundamental physiological processes. Perhaps the most important example is the work on the phototropic growth of grass coleoptiles which led to the discovery of indoleacetic acid (IAA), the first known plant hormone. The concept of hormones, which originated in the field of animal physiology, was employed by botanists for the first time to explain these responses and revolutionized plant physiology.

A number of kinds of plant responses to environmental signals will be demonstrated in this laboratory. For gravitropism and phototropism, you will attempt to parallel the kinds of experiments and reasoning which first led Charles Darwin and other scientists to propose the existence of plant hormones. The electrical changes that are associated with

157

rapid plant movements are the subject of current study; Part VI will consist of a demonstration of action potentials associated with the thigmonastic responses of *Mimosa* and Venus flytrap plants.

MATERIALS

Equipment
single edge razor blades
aluminum foil
applicator sticks
lanolin
blue, green and red light filters
black paper tents
single direction light source
silver wire
Elmer's glue
ice water
Pasteur pipet and bulb
1 mV recorder

Solutions
1% naphthaleneacetic acid (NAA) in lanolin
1% triiodobenzoic acid (TIBA) in lanolin

Plant material
sunflower (*Helianthus annuus* var.
    Sungold) plants, 5 weeks old
oat (*Avena sativa* var. Porter) seedlings,
    6 days old, dark grown
pea (*Pisum sativum* var. Progress #9)
    plants, 3 weeks old
Scarlet Runner bean (*Phaseolus coccineus*),
    4 weeks old, wound clockwise and counter-
    clockwise on supports
Red Kidney bean (*Phaseolus vulgaris* var.
    Redcloud) plants, 3 weeks old
Oxalis spp. plants
Mimosa pudica plant
Venus flytrap plant (*Dionaea muscipula*) plant

PART I. Gravitropism.

Plant growth is classified as tropistic when it consists of a directional response to a stimulus which is determined by the direction from which the stimulus is received. **Gravitropism** is the directional growth response of a plant to the earth's gravitational field. For most roots, the response is in the direction of the gravitational vector and is thus an example of **positive gravitropism**; shoots, which bend away from the earth, are said to exhibit **negative gravitropism**.

The response of stems to gravity is thought to be mediated by indoleacetic acid (IAA) produced at the apex. IAA (or "auxin") travels **down** the stem (called **polar movement**) and promotes growth in the area just below the apex. One model for gravitropic response suggests that when a plant is placed horizontally, IAA accumulates on the lower side and leads to an increase in the growth of the cells on that side, resulting in an upward bending of the stem.

You will place sunflower plants in a horizontal position and observe the gravitropic response. By removing the apex in one treatment and replacing it with the synthetic auxin naphthaleneacetic acid (NAA) in another treatment, you will test the model suggested above. The compound triiodobenzoic acid (TIBA), which inhibits the polar movement of auxin, will be applied in some treatments. All experimental treatments will be performed in the dark to prevent interference from phototropic growth.

PROCEDURE

(1)   Label five pots of sunflower plants for the treatments below. Treat each plant as directed.

   1.   Whole plant; place horizontally.
   2.   Decapitate plant; apply lanolin to cut end, place horizontally.
   3.   Decapitate plant; apply 1% NAA in lanolin to the cut end, place horizontally.
   4.   Whole plant; apply a ring of 1% TIBA in lanolin just below the apex and place horizontally.
   5.   Decapitate plant; apply 1% NAA in lanolin to the cut end and a ring of 1% TIBA in lanolin just below the apex and place horizontally.

(2)   Place all of the plants on their sides in the dark. Mark one side of each pot and make sure that it remains in exactly the same orientation. Record changes in the direction of stem growth every 30 minutes over the next two hours. If no changes are noted during this time, examine the plants after three hours and again the next morning. Record your observations by simple sketches in the chart on page 163 and classify bending for each treatment as 0, +, + +, or + + +.

PART II. Phototropism.

The directional growth of plants in response to a light stimulus, or **phototropism**, will be explored in the following experiment, which is designed along the lines of one of Darwin's original experiments, found in his book on the power of movement in plants. Oat seedlings which have been grown in the dark to develop elongate coleoptiles will be provided. Figure 18-1 gives a model for the auxin control of the phototropic response.

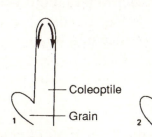

FIGURE 18-1. A model for auxin control of the phototropic response. (Adapted from Galston, Davies & Satter, *The Life of the Green Plant*, p. 222, Prentice Hall, 1980.)

A coleoptile growing in the dark or in symmetrical light has symmetrical auxin distribution and symmetrical growth.

A coleoptile exposed to unilateral light develops an asymmetry in auxin distribution.

This leads to an asymmetry in growth which results in curvature.

PROCEDURE

(1)   Keeping your container of oat seedlings in dim light only, place small caps of aluminum foil (made on a pencil tip) over the tops of four coleoptiles. The caps should be about 3 mm long. Place a 1-cm collar made from one layer of aluminum foil just below the tip of four additional coleoptiles. Cut the top 3 mm from four more coleoptiles and leave the rest untouched. Surround the container on all but one side with a black paper tent.

(2)   If color filters and light-tight boxes are available, place additional containers of untreated oat seedlings in a light-tight box which provides unidirectional light passed through color filters to determine the effects of blue, green and red light on the phototropic response.

(3)   Illuminate all of the coleoptiles from one side for two to three hours. Note when they bend, which ones do so, and the site of the bending. Record your observations in the chart on page 164. Determine which color of light is most effective in stimulating a positive phototropic response.

PART III. Thigmotropism.

Thigmotropism is the directional growth of plants in response to touch; climbing plants respond to touching a support by twining around it. The response of pea tendrils to touch has been classified as thigmotropic by some and thigmonastic by others. Try the procedure described below on two different tendrils, stroking the upper side of the tendril in one case, and the lower side in the other.

PROCEDURE

With a glass rod, gently stroke the upper or lower side of a well-developed tendril of a pea plant five or six times. Check the orientation of the tendril every few minutes for the next 10 to 15 minutes. Record your observations with simple sketches in the chart on page 164 and note how long it takes before bending is observed. You may find that the tendril responds quite quickly to the presence of a "support."

PART IV. Nutational Movements.

The stems of many plants grow with pendulum-like or spiraling motions called **nutation**; in some climbing plants, nutation results in twining. Examine the two Scarlet Runner bean plants (one was wrapped clockwise around a support and the other was wrapped counterclockwise) and note the pattern of growth below and above the point where the plants were tied to the support. Nutational movements are reduced when plants are grown in a clinostat, indicating that the sensing of gravity may be involved in the control of these movements. Include a description of these plants in the chart on page 164.

PART V. Nyctinasty.

The leaves of plants in some plant families fold at night and open during the day. Beans and *Oxalis* both show pronounced nyctinastic leaf movements. Examples of **nyctinastic** or "sleep" movements may be observed in plants kept in an incubator which provides light at night and darkness during the day. Note the position of the leaves of the *Oxalis* plants in the incubator and compare it with the leaf position of the leaves of a plant kept in room light. Make the same observations on the primary leaves of a bean plant. Make a sketch of each plant in the chart on page 165. These movements are classified as nastic because the plant's response is unaffected by the direction from which the controlling stimulus (light) is received. The nyctinastic movements of *Albizzia* and *Mimosa* are known to be influenced by phytochrome (see Expt. 26).

PART VI. Thigmonasty Demonstration.

Many plants are able to sense mechanical stimuli and respond to them. In some carnivorous plants these responses have reached a level of sophistication which surpasses that of a few lower animals. *Dionaea* (the Venus flytrap) has a rapid thigmonastic response which is mediated by an action potential (the type of signal which passes down animal nerves). These action potentials are initiated by a receptor potential which occurs in the sensory receptor organs, the hairs on the upper surface of the traps in the Venus fly trap. The receptor potentials behave in a manner similar to those in the sensory receptors of animals. It is likely that most rapid nastic responses in plants are action potential mediated.

The instructor will attach a fine silver wire (input electrode) to the upper surface of a trap with some Elmer's glue and place another silver wire in the soil as a reference electrode. Both are connected to a recorder or millivoltmeter, so that the electrical changes associated with rapid movements may be recorded. When the hairs on the upper surface are stimulated, an action potential should be observed and the trap should close.

The leaves of the *Mimosa* plant bend downward at the base of the petiole in response to touch. By attaching the input electrode to the surface of the petiole with some Elmer's glue (midway between the leaflets and the base of the petiole) and placing the reference electrode in the soil, you may observe an action potential associated with the bending of the leaf. The leaf may be stimulated by dispensing several drops of ice water onto the petiole near the base of the leaflet blades. As the electrical signal travels down the petiole an action potential should be observed, followed by the dropping of the petiole at its point of attachment to the stem.

TEXT REFERENCES

Galston, A. W., P. J. Davies, and R. L. Satter, *The Life of the Green Plant* (3rd ed.), pp. 222-229 and Chap. 13. Englewood Cliffs, NJ: Prentice Hall, 1980.

Leopold A. C. and P. E. Kriedman, *Plant Growth and Development* (2nd ed.), pp. 207-215. New York: McGraw-Hill, Inc., 1975.

Salisbury, F. T. and C. W. Ross, *Plant Physiology* (4th ed.), Chap. 19. Belmont, CA: Wadsworth Publishing Co., 1991.

Taiz, L. and E. Zeiger, *Plant Physiology*, pp. 411-415. Redwood City, CA: Benjamin/Cummings Publishing Co., Inc., 1991.

FURTHER READING

Bandurski, R. S., A. Schulze, P. Dayanandan and P. B. Kaufman, "Response to gravity by *Zea mays* seedlings. I. Time course of the response," *Plant Physiol.* (1984), 74:284-288.

Baskin, T. I., W. R. Briggs and M. Iino, "Can lateral redistribution of auxin account for phototropism in maize coleoptiles?" *Plant Physiol.* (1986), 81:306-309.

Britz, S. J. and A. W. Galston, "Physiology of movements in the stems of seedling *Pisum sativum* L. cv Alaska. III. Phototropism in relation to gravitropism, nutation, and growth," *Plant Physiol.* (1983), 71:313-318.

Brown, A. H. and D. K. Chapman, "Circumnutation observed without significant gravitational force in spaceflight," *Science* (1984), 225:230-232.

Brown, A. H. and D. K. Chapman, "Kinetics of suppression of circumnutation by clinostatting favors modified internal oscillator model," *Amer. J. Bot.* (1988), 75(8):1247-1251.

Darwin, C. (assisted by F. Darwin), *The Power of Movement in Plants*. New York: Appleton-Century-Crofts, 1881.

Evans, M. L., "Rapid responses to plant hormones," *Annu. Rev. Plant Physiol.* (1974), 25:195-223.

Harrison, M. A. and B. G. Pickard, "Auxin asymmetry during gravitropism by tomato hypocotyls," *Plant Physiol.* (1989), 89:652-657.

Hart, J. W., *Plant Tropisms and Other Growth movements*. London: Unwin Hyman, 1990.

Hodick, D. and A. Sievers, "On the mechanism of trap closure of Venus flytrap (*Dionaea muscipula* Ellis)," *Planta* (1989), 179:32-42.

Pickard, B. G., "Action potentials in higher plants," *Botanical Review* (1973), 39:172-201.

Pickard, B. G., "Roles of hormones, protons and calcium in geotropism," pp. 193-281 in R.P. Pharis and D.M. Reid, eds., *Encyclopedia of plant physiology (New Series)*, Vol. 11, *Hormonal Regulation of Development III*. Berlin: Springer-Verlag, 1985.

Williams, S. and A. B. Bennett, "Leaf closure in the Venus flytrap: an acid growth response," *Science* (1982), 218:1120-1122.

EXPERIMENT 18

RESULTS

Name _____
Date _____

Record your observations in the charts that follow by sketches and careful descriptions.

PART I. Gravitropism. Record changes in the growth of the stems by sketches the classify the degree of bending observed as 0, +, + +, or + + +.

Gravitropism: sketches of stems.						
Time:	0 min	30 min	60 min	90 min	120 min	Bending?
Control						
Decapitated						
Decapitated + NAA						
+TIBA						
Decapitated + NAA +TIBA						

PART II. Phototropism. Record bending as 0, +, + +, or + + +.

Treatment	Bending?
Control	
+ Cap	
Decapitated	
+ Collar	

Treatment	Bending?
Blue	
Green	
Red	

PART III. Thigmotropism. Make sketches of the bending observed in tendrils when they are stroked on the upper and lower sides.

Time:					
Upper Side					
Lower Side					

PART IV. Nutational movements. Describe your observations of stem growth.

Treatment	Description
Clockwise	
Counterclockwise	

PART V. Nyctinasty. Make drawings of the plants you observed. Note the position of the petioles and leaf blades.

Treatment	Bean	*Oxalis*
Light		
Dark		

PART VI. Thigmonasty. Record your observations in the space below.

QUESTIONS

1.   What is the difference between a tropism and a nastic movement?

2.   How does the plant react relative to the gravitational stimulus? Why were the plants kept in the dark for the gravitropism experiment?

3.   Can you determine at what site the gravitational stimulus is perceived? How might auxin mediate this response? What evidence do you have that auxin is the mediator?

4.   At what site is the phototropic stimulus perceived? What evidence do you have that the effect is a transmitted one?

5.   What colors of visible light are most effective in stimulating a phototropic response? What pigment is most likely to be the photoreceptor?

6.   Why might the response of pea tendrils to touch be classified as thigmonastic **and** thigmotropic.

7.   Briefly discuss nutational movements, nyctinasty and thigmonasty, indicating what information about movement in plants each experiment or demonstration provided.

EXPERIMENT 19

Calcium and Signal Transduction: Cytoplasmic Streaming in *Chara* Cells

INTRODUCTION

Some plant cells exhibit cytoplasmic streaming, which accelerates the transport and mixing of substances in the cell; streaming is particularly important in large cells such as found in *Chara*, a fresh water alga. The movement of the cytoplasm is controlled by an actin and myosin system which requires ATP and is similar in some ways to the system used for muscle movement in animals.

Cytoplasmic streaming in a *Chara* cell can be induced to stop instantly if the membrane is mechanically or electrically excited. When the cell is stimulated, an action potential (or change in membrane potential associated with a response) is created. The action potential moves from one end of the cell to the other and leads to the opening of Ca^{2+} channels, allowing calcium ions to flow from the cell wall (or the surrounding medium) inward to the cytoplasm along an electrochemical potential gradient. As has been observed in animal systems, a temporary increase in the cytoplasmic Ca^{2+} concentration can lead to a cascade of events; hence, Ca^{2+} is said to be a crucial element in a signal transduction system. The signal transduction chain is a series of events which begins when a cell is stimulated and Ca^{2+} channels open (in some cases because of a change in membrane potential). The resulting increase in cytoplasmic Ca^{2+} may trigger a biochemical change (for instance, in the activity of an enzyme), which may produce another biochemical change, which may have yet further effects. In the case of the cessation of cytoplasmic streaming, Ca^{2+} acts by activating a protein kinase, which leads to the phosphorylation of a protein, possibly myosin. Micromolar external concentrations of Ca^{2+} are required.

In this experiment, you will observe cytoplasmic streaming in *Chara* cells and attempt to stop streaming by producing an action potential in the presence of an external concentration of 10^{-4} M Ca^{2+} (or pCa 4, using the same notation as for pH). To determine if Ca^{2+} is required for the response, the experiment will be repeated with a lower external concentration of Ca^{2+} (pCa 7 or 10^{-7} M Ca^{2+}).

MATERIALS

Equipment
microscope (low power)
reaction chamber
silicone grease or petroleum jelly
transfer pipets
1.5 V battery (AAA)
tape
plastic petri dishes
Kimwipes

Solutions
10^{-4} M $CaCl_2$, buffered with 1 mM EGTA and
    2 mM HEPES, 0.1 mM NaCl and 0.1 mM KCl
10^{-7} M $CaCl_2$, buffered with 1 mM EGTA and
    2 mM HEPES, 0.1 mM NaCl and 0.1 mM KCl
10 mM EGTA

Plant Material
Chara cells, harvested and floating in pCa 4 buffer

PROCEDURE

(1)   Choose a *Chara* cell from the petri dish containing cells in pCa 4 solution. Carefully lift the cell, using forceps to grasp the cell gently **at one end**. Quickly and gently blot the cell on toweling or a Kimwipe, then transfer it to a petri dish containing the pCa 7 solution. Let the cell incubate for at least 20 minutes. Continue with step (2).

(2)   Select another *Chara* cell from the pCa 4 solution (grasp it at the end only) and blot the cell carefully on a Kimwipe to remove excess solution which might provide a low resistance pathway for the applied electrical charge. Gently transfer the cell to the specially constructed glass slide or "reaction chamber." The reaction chamber contains two reservoirs separated by a wall of silicon grease or petroleum jelly. The chamber is designed so that the middle of the rather long *Chara* cell rests in grease or petroleum jelly, and each end rests in a separate reservoir of solution. Line up your cell properly, adding additional grease at the center of the cell if necessary, and add several drops of the pCa 4 solution to each reservoir. It is important that no liquid conduit exist between the reservoirs, so blot up any excess solution.

(3)   Place the reaction chamber on the microscope stage and examine the cell under low power. Look for a clear strip between rows of chloroplasts. Notice that cytoplasmic streaming may be observed in the areas associated with the chloroplasts and that streaming occurs in two directions. If you do not observe streaming, focus up and down slightly until streaming is seen. If you still cannot see any movement of particles, wait several minutes; the process of transferring the cell may have set off an action potential and stopped cytoplasmic streaming. The cell should recover

within a few minutes; if it does not, try a new cell and be **gentle**. Once you have found a suitable site for your observations of streaming, secure the slide in place on the microscope stage.

(4)   Each reservoir is in contact with a slender wire which extends out of the reservoir and, when both wires are connected to a battery, will allow you to create an electrical current across the cell. Tape a 1.5 V battery to the edge of the microscope stage, then tape one of the wires to the negative (flat) end of the battery.

(5)   **While you are looking at the streaming cell** under the microscope, touch the unattached wire to the positive end of the battery. Your lab partner may be able to help by making the connection and saying "now." Record your observations in the chart on page 171. Observe the cell at 15 second intervals for the next minute or so. When streaming has resumed, the action potential may be repeated for the benefit of your lab partner.

(6)   Repeat steps (2) through (5), using the cell which has been stored in the pCa 7 solution for 20 minutes. Rinse the reaction chamber with EGTA before starting. Record your observations in the chart on page 171.

TEXT REFERENCES

Salisbury, F. B. and C. W. Ross, *Plant Physiology* (4th ed.), pp 408-416. Belmont, CA: Wadsworth Publishing Co., 1991.

Taiz, L. and E. Zeiger, *Plant Physiology*, pp. 417-418 and 505-507. Redwood City, CA: Benjamin/Cummings Publishing Co., Inc., 1991.

FURTHER READING

Shimmen, T. and M. Tazawa, "Cytoplasmic streaming in the cell model of *Nitella*," *Protoplasma* (1982), 112:101-106.

Tominaga, Y. and M. Tazawa, "Reversible inhibition of cytoplasmic streaming by intracellular Ca^{2+} in tonoplast-free cells of *Chara australis*," *Protoplasma* (1981), 109: 103-111.

Wayne, R., M. Staves, and A. C. Leopold, "The influence of gravity on the polarity of cytoplasmic streaming in *Nitellopsis*," *Protoplasma* (1990), 155:43-57.

EXPERIMENT 19   Name _____
                                                        Date _____

RESULTS

Record your observations in the chart below:

Solution	Observations of Streaming

QUESTIONS

1.   What is an action potential?

2.   Did your results indicate that Ca^{2+} is required for the cessation of cytoplasmic streaming? Give an explanation of the role of Ca^{2+} in this experiment.

EXPERIMENT 20

Apical Dominance

INTRODUCTION

In many plants the lateral buds do not develop when the terminal meristem is active. The control of lateral bud growth by the apical meristem is called apical dominance; when the apex is removed the lateral buds grow out and form branches. This concept has long been understood by gardeners, who pinch back the growing tip to produce a bushier plant. One model for this phenomenon suggests that auxin moving down the stem from the apex prevents the outgrowth of buds in the axils of leaves near the apex. Cytokinins, moving up the stem in the transpiration stream from the roots, can enhance the growth of lateral buds.

In the following experiment, you will attempt to demonstrate that the terminal meristem may be replaced by an artificial source of auxin (naphthaleneacetic acid will be used because it is more stable in light than indoleacetic acid). In some experimental treatments, benzylaminopurine, a cytokinin, will be applied to the lateral buds to explore the possibility of an interaction of auxin and cytokinin effects. The kinds of studies performed in this experiment are called replacement studies. A plant part, such as the apex, is removed and replaced with a hormone to answer the question, "Can the hormone **replace** the removed organ and produce the **same effect**?" If the answer is yes, the results are considered good evidence that the hormone is involved in the phenomenon. Such studies are considered to be a **first step** in demonstrating that a hormone is the controlling factor in a particular system. Further studies would be required before the hormone is considered to be the causal agent. The hormone must be extracted from the organ (to show that it is present), then isolated and identified. If the extracted hormone, when reapplied, acts to replace the removed organ, then the hormone is considered to be of primary importance in the phenomenon.

MATERIALS

Equipment
razor blades
applicator sticks
small paint brush
space in a greenhouse
trays large enough to hold 5 pots

Solutions
1% (by weight) naphthaleneacetic
    acid (NAA) in lanolin
1.5 mM 6-benzylaminopurine (BA)
    + 0.1% Tween 20 in lanolin

Plant material
Red Kidney bean (*Phaseolus vulgaris* var. Redcloud) plants, 3 weeks old, 2 plants per pot

---

PROCEDURE

Plants will be treated with an auxin and/or a cytokinin and the length of the lateral buds determined. The plants will be placed in the greenhouse for two weeks and four additional measurements of the lateral bud length will be made.

(1) Five pots will be provided, each containing two bean plants. Label each pot according to the list of treatments below. Use two plants (four lateral buds) per treatment. Refer to Appendix F if you are unsure which are the primary leaves.

   1. Control. Leave the plant intact.
   2. Cut off the terminal part of the stem about 1 cm above the primary node, leaving one pair of primary leaves (which have suppressed buds in their leaf axils) on each plant. Examine the buds to be sure you know what they look like. Apply lanolin to the cut end of the stem.
   3. Cut off the top as in 2; but smear the cut surface with lanolin containing 1% naphthaleneacetic acid (NAA), an artificial auxin.
   4. Leave plant intact, but apply 1.5 mM benzylaminopurine (BA), which is a cytokinin, to the lateral buds of the primary node by painting it on the surface. Apply a second treatment the following day.
   5. Cut off the top and apply NAA (as in 3) but also apply benzylaminopurine to the lateral buds (as in 4). Apply a second treatment of BA the following day.

(2) Measure the length of the primary lateral buds (those in the axils of the primary leaves) for each treatment and record your measurements on the chart on page 177.

(3) Place the plants in a tray and put the tray in the greenhouse. Twice a week, measure the length of the primary lateral buds; find the average for each treatment. Continue measurements for two weeks. Finally, plot the bud length against time on the graph on page 178.

TEXT REFERENCES

Galston, A. W., P. J. Davies, and R. L. Satter, *The Life of the Green Plant* (3rd ed.), p. 249. Englewood Cliffs, NJ: Prentice Hall, 1980.

Salisbury, F. B. and C. W. Ross, *Plant Physiology* (4th ed.), pp. 368-369 and 388-389. Belmont, CA: Wadsworth Publishing Co., 1991.

Taiz, L. and E. Zeiger, *Plant Physiology*, pp. 415-6. Redwood City, CA: Benjamin/Cummings Publishing Co., Inc., 1991.

FURTHER READING

Phillips, I. D. J., "Apical Dominance," *Annu. Rev. Plant Physiol.* (1975), 26:341-367.

Rubenstein, B. and M. A. Nagao, "Lateral bud outgrowth and its control by the apex," *Bot. Rev.* (1976), 42:83-109.

EXPERIMENT 20

Name _____

Date _____

RESULTS

Record your data in the chart below. Determine the average bud length at each time period.

Lateral Bud Length (cm)					
Date	Control	Decapitated	Decapitated + NAA	BA on Buds	Decapitated + NAA + BA
Initial Length	—— ——	—— ——	—— ——	—— ——	—— ——
Average Bud Length (cm):	—— ——	—— ——	—— ——	—— ——	—— ——
	—— ——	—— ——	—— ——	—— ——	—— ——
Average Bud Length (cm):	—— ——	—— ——	—— ——	—— ——	—— ——
	—— ——	—— ——	—— ——	—— ——	—— ——
Average Bud Length (cm):	—— ——	—— ——	—— ——	—— ——	—— ——
	—— ——	—— ——	—— ——	—— ——	—— ——
Average Bud Length (cm):	—— ——	—— ——	—— ——	—— ——	—— ——
	—— ——	—— ——	—— ——	—— ——	—— ——
Average Bud Length (cm):	—— ——	—— ——	—— ——	—— ——	—— ——

Plot the average lateral bud length as a function of time in the graph below.

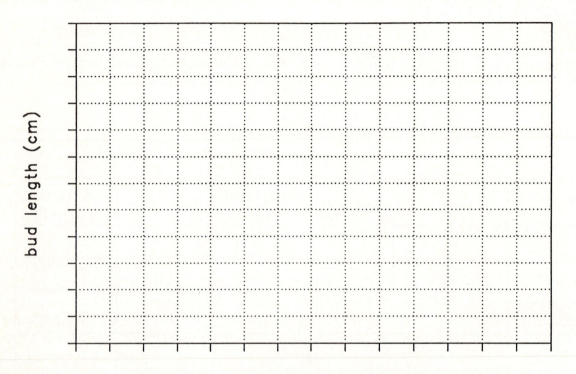

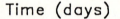

Time (days)

QUESTIONS

1. As you know, one effect of auxin is to increase cell wall extension in certain types of cells. In explaining auxin's role in apical dominance, it has been suggested that nutrient diversion away from lateral buds could be brought about by auxin located in the apical bud. Considering the fundamental effect of auxin, what does it mean when it is said that auxin "attracts" nutrients away from the lateral buds?

2. What is the meaning of the following statement: There is a "trade-off" between auxin and cytokinin effects in apical dominance and lateral bud outgrowth. A diagram might be a suitable way to discuss this statement.

3. Consider the following information. A particular fern, *Microgramma*, displays strict apical dominance. When its shoot tip (apical bud) is removed, the lateral buds grow out. Replacement of the shoot tip with indoleacetic acid (IAA) preserves apical dominance. However, biochemical tests (extraction and diffusion techniques followed by chromatography) failed to reveal any endogenous auxin activity in the plant. Assuming the accuracy of the above results, what does this information tell you about the usefulness of the experimental protocol we used in our lab in determining the involvement of auxin in apical dominance?

EXPERIMENT 21

Leaf Senescence

INTRODUCTION

Leaf senescence is most easily recognized by the loss of green color associated with the breakdown of chlorophyll and the chloroplasts. The physiological events (such as loss of membrane integrity and a decline in mRNA and protein synthesis) that occur during senescence are a controlled series of steps leading to the death of the leaf. Although the events associated with senescence in leaves are well characterized, the biochemical changes that trigger the onset of senescence are still the object of study. A variety of factors, both internal and external, may bring on leaf senescence. In deciduous trees, the leaves lose their green color with the short photoperiod and cool temperatures of autumn, indicating that senescence may be triggered by environmental cues. In some plants the leaves senesce when the fruits mature, which implies that there is competition for some factor which delays senescence. A shortage of mobile nutrients combined with competition from younger leaves will make older leaves senesce, regardless of the age of the whole plant. Detached leaves senesce more rapidly than attached leaves of the same age, but detached leaves that go on to produce roots may survive indefinitely.

The examples cited above all imply that hormonal interactions control the onset of senescence. Cytokinins (e.g., benzylaminopurine and kinetin) have long been known to delay senescence in leaves; more recently, polyamines like spermidine have been shown to slow the onset of senescence as well. Both may act by maintaining membrane integrity; both are also known to enhance protein synthesis. Their exact biochemical role in senescence has yet to be determined. You will perform several simple experiments to test the concept that cytokinins and competition for nutrients and hormones are involved in controlling the onset of senescence.

MATERIALS

PART I

Equipment	**Solutions**
razor blades	0.2 mM 6-benzylaminopurine
Munsell color chart, 2.5 GY and/or 5 GY	

Plant material
Red Kidney bean (*Phaseolus vulgaris* var. Redcloud), 4-6 weeks old

PARTS II and III

Equipment	**Solutions**
24-mL shell vials	0.1 mM kinetin
Parafilm	(1 mM spermidine)
Munsell color chart, 2.5 GY and/or 5 GY	
(9-cm plastic petri dishes)	

Plant material
wheat (*Triticum aestivum* var. Frankenmuth) plants, 7 days old

---

PART I. Senescence of Leaves on Whole Bean Plants.

Whole plants will be treated then placed in the greenhouse for two weeks. The bean plants you will use have been grown under conditions of suboptimal nutrition so that the oldest leaves have just a trace of yellowish cast appearing. The color of the primary leaves (see Appendix F) should be classified using a Munsell color chart at the outset of the experiment; four additional observations of color in the primary leaves should be made.

PROCEDURE

(1)   Use four bean plants (one with developing fruit) and treat the two oldest leaves (the primary leaves) as follows:

    1.   Control. No treatment.

    2.   Paint the primary leaves with 0.2 mM benzylaminopurine or BA (a cytokinin).

    3.   Leave the primary leaves intact as in the control. Remove all other leaves and growing points. Remove any branchlets which appear during the course of the experiment.

    4.   The primary leaves of a plant with developing fruits should be observed until the fruits mature. The three treatments above should be done with plants that are still vegetative.

(2)   Classify the color of the primary leaves using a Munsell color chart. Make observations of color changes in the primary leaves twice a week for the next two

weeks. Refer to the Munsell color chart to classify the color each time, and determine the number of days to senescence (senescence can be defined as a color agreed upon by the class). Record the color changes (as Munsell color numbers) in the chart on page 183.

PART II. Senescence of Detached Leaves of Wheat.

Wheat leaves will be placed in shell vials which contain water or kinetin; changes in leaf color will be noted every two days for two weeks.

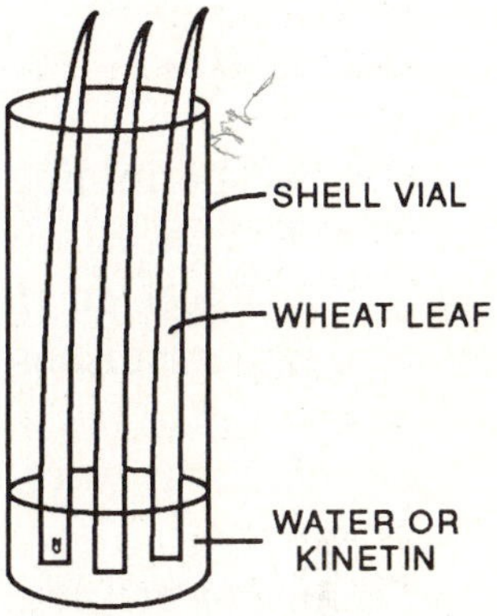

PROCEDURE

(1)   Label two shell vials for the treatments below and add 5 mL of the appropriate solution to each vial.

  a) Distilled water
  b) 0.1 mM kinetin

FIGURE 21-1. Detached wheat leaves placed in solution in a shell vial.

(2)   Cut six wheat leaves to about 5 cm in length and stand three leaves in each vial, with the cut end of the leaves placed in the liquid (Fig. 21-1). Store the vials in the dark at room temperature. Be sure to replenish the solution with water or kinetin (as appropriate to the treatment) if you notice a significant drop in the level of liquid.

The leaves kept in water should begin to show signs of senescence within a few days; those kept in kinetin may remain green for at least that length of time. Examine the leaves every two days for the next two weeks. Refer to the Munsell color chart to classify the leaf color and determine the number of days to senescence (senescence can be defined as a color agreed upon by the class). Use the chart for Part II on page 183.

PART III. Design Your Own Experiment about Leaf Senescence.

You are asked to design an experiment to answer the following question (or one of those listed on the next page): Can cytokinin rescue an aging detached leaf once it has started to senesce? Start an additional set of four or five vials with detached leaves in distilled water and perform an appropriate experiment to answer the question above. If you choose wheat leaves as your plant material, you may use the treatments from Part II as your controls. Use the space on page 184 to describe your experimental protocol.

OR, determine what concentration of kinetin is required to delay senescence.

OR, if you have a particular interest in some other plant whose leaves are available, design an experiment to test for the effect of cytokinin or another hormone on that plant in place of doing the above experiment.

OR, if you are interested, determine the effect of light on the rate of senescence with added kinetin.

OR, as spermidine has been shown to delay senescence, you might want to test its effect on senescence in wheat leaves. Since spermidine is not readily transported within the leaf, it is suggested that several 1-cm sections of wheat leaf be floated in a solution of spermidine in petri dishes kept in the dark.

OR, you may want to test for the effect of light on spermidine-delayed senescence.

Be sure to prepare the proper controls. Ask your instructor for assistance if you are unsure what treatments are needed to answer your chosen question.

TEXT REFERENCES

Galston, A. W., P. J. Davies, and R. L. Satter, *The Life of the Green Plant* (3rd ed.), pp. 260-269. Englewood Cliffs, NJ: Prentice Hall, 1980.

Salisbury, F. B. and C. W. Ross, *Plant Physiology* (4th ed.), pp. 334-5. Belmont, CA: Wadsworth Publishing Co., 1991..

Taiz, L. and E. Zeiger, *Plant Physiology*, p. 465. Redwood City, CA: Benjamin/Cummings Publishing Co., Inc., 1991.

FURTHER READING

Evans, P. T. and R. L. Malmberg, "Do polyamines have roles in plant development?" *Annu. Rev. Plant Physiol. Plant Mol. Bio.* (1989), 40:235-269.

Kelly, M. O. and P. J. Davies, "The control of whole plant senescence," *CRC Critical Reviews in Plant Sciences* (1988), 7:139-173.

Leopold, A. C., and M. Kawase, "Benzyladenine effects on bean leaf growth and senescence," *Am. J. Bot.* (1964), 51:294-298.

Nooden, L. D., and A. C. Leopold, eds., *Senescence and Aging in Plants*. San Diego: Academic Press, 1988.

Thimann, K. V., ed., *Senescence in Plants*. Boca Raton, FL: CRC Press, 1980.

EXPERIMENT 21

Name _____

Date _____

RESULTS

PART I.

Record your data in the chart below and determine the number of days to senescence for each treatment.

#	Date:	Senescence: color changes					Days to Senescence
1	Control						
2	+ BA						
3	Remove other Leaves						
4	+ Fruit						

PART II.

Record your data in the chart below and determine the number of days to senescence for each treatment.

	Senescence: color changes								
Date:									
Control									
Kinetin									

PART III.

Experimental protocol:

Record your results and conclusions here.

EXPERIMENT 22

Hormones and Leaf Abscission

INTRODUCTION

It is well known that most plants lose their leaves once they have senesced. The mechanism for leaf abscission involves formation or activation of a specialized abscission zone, a layer of cells close to the base of the petiole (Fig. 22-1). Leaf abscission results from a complex series of hormonally controlled events, although other factors can have a strong influence on the speed with which the abscission layer forms. In young leaves, auxin is synthesized in the leaf blade and transported down the petiole. The constant supply of auxin maintains the metabolic status of the petiole and prevents abscission; removal of the leaf blade will stimulate abscission. As the leaf ages, the auxin production and transport declines, triggering changes in the abscission zone. Certainly the age of the petiole is important; applied auxin may delay senescence, but, if the petiole tissues have aged, may speed the time to abscission by encouraging the production of ethylene there. In senescing leaves, newly synthesized ethylene in the abscission layer promotes the synthesis of enzymes, such as cellulase and pectinase, which degrade the cell walls of the abscission zone. Ethylene also controls the release of these enzymes

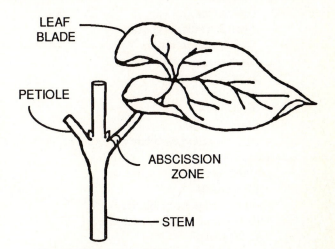

FIGURE 22-1. The position of one of the abscission zones on the petiole of a primary leaf of a bean plant. A second abscission zone is located at the base of the leaf blade.

into the cell wall and induces swelling in the abscission zone cells closest to the plant stem. According to the suggested model, then, abscission is an active, hormonally controlled process which is dependent on the age of the tissue and the presence of auxin and ethylene.

In Part I, you will observe leaf abscission in whole plants and perform a replacement study to determine if applied auxin can take the place of the leaf blade in maintaining the status of the petiole and preventing abscission. In Part II, you will use excised portions (explants) of *Albizzia* plants. Explants age more rapidly than attached tissues or whole plants. Both auxin and ethylene will be applied and the time to abscission determined. Cycloheximide, a protein synthesis inhibitor, will be added to determine if protein synthesis is required for abscission, as suggested by the model for active control of leaf abscission. In Part III, tissue prints of abscission zones will be made and stained for protein content.

MATERIALS

PART I
Equipment   **Solutions**
single edge razor blades   1% naphthaleneacetic acid (NAA) in lanolin
applicator sticks   lanolin

Plant material
Red kidney bean (*Phaseolus vulgaris* var. Redcloud) plants, 4 weeks old, two per pot

PART II
Equipment   **Solutions**
9-cm petri dishes containing $1\,\mu g/mL$ (1 ppm) cycloheximide (CHI)
   3% agar in water   in lanolin
hot plate   0.2 mL/L (200 ppm) ethephon
9-cm filter paper   0.05 M phosphate buffer, pH 7.5
Parafilm   1% naphthaleneacetic acid (NAA) in lanolin
                                                    lanolin

Plant material
Albizzia julibrissin plants or other plants with compound leaves

PART III
Equipment   **Solutions**
double edge razor blades   0.05% CPTS stain for protein
$0.45\text{-}\mu$ m nitrocellulose sheets   in 12 mM HCl
microscope   12 mM HCl
glass slides
forceps

Plant material
Red kidney bean (*Phaseolus vulgaris* var. Redcloud) plants, 4 weeks old, with and without leaf blades

PART I. Abscission of Attached Petioles of Bean Plants.

Whole bean plants will be treated and kept in the greenhouse for two weeks. After one week, the number of abscised leaves will be counted daily to determine the time until 50% of the primary leaves have abscised.

PROCEDURE

(1)   Select three pots of plants, with two plants per pot. Use two plants (four petioles) for each of the treatments that follow:

      1.   Control. Keep the plants intact.
      2.   Bladeless petioles. Cut off the leaf blades of the primary leaves, leaving about 1 cm of petiole intact. Apply a dab of pure lanolin to the cut end.
      3.   Bladeless petioles + auxin. Remove the leaf as in treatment 2, but apply a dab of 1% naphthaleneacetic acid (NAA) in lanolin, to the cut end.

(2)   Place the plants in the greenhouse. After one week, examine the plants daily and count the number of abscised leaves. Use the chart on page 193 to record your data. At the end of two weeks, determine the number of days until 50% abscission.

PART II. Abscission in Explants from Compound Leaves.

Explants may be any portion of a plant or plant organ. For this experiment, the explant consists of a small section of rachilla, pulvinule (containing the abscission zone) and leaflet (Fig. 22-2), but explants could just as well be composed of a small section of petiole and rachilla. *Albizzia* leaflets will be available for study, but if you are interested in trying another species, please do so.

PROCEDURE

(1)   Remove a strip of agar about 2 cm wide from the center of five petri dishes, leaving a narrow bridge of agar across one end to hold the agar in place.

(2)   Remove a section of *Albizzia* leaf from the stem. Cut the rachilla into sections containing three pinnule pairs (six leaflets), then remove the lower two pinnule pair leaflets from the rachilla. Finally, making a diagonal cut, trim away much of the pinnule blade from the main vein of the leaflet (Fig. 22-2).

(3)   Place six explants in each petri dish with the rachilla embedded in the agar and the abscission zone overhanging the central channel.

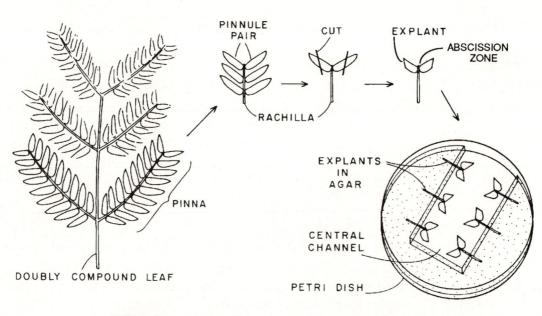

FIGURE 22-2. Explants from a doubly compound leaf.

(4)   Some hormones and inhibitors may be applied to the explants in a lanolin paste. For the treatments 2, 4 and 5 below, add the appropriate lanolin mixture (warmed until the lanolin is syrupy) with an applicator stick. One or two drops of the lanolin mixture should be applied directly to the abscission zone. NAA (naphthaleneacetic acid) is an analogue to IAA. CHI (cycloheximide) is a protein synthesis inhibitor.

> 1.   Control; no treatment.
> 2.   + NAA in lanolin.
> 3.   + ethephon.
> 4.   + CHI in lanolin.
> 5.   + CHI in lanolin + ethephon.

Since ethylene is a gas, application must be accomplished with a different technique. If a piece of filter paper is placed on the inside of the lid and inoculated with an ethylene-generating system, sufficient ethylene to affect the rate of abscission will be produced. Ethylene may be generated by pipetting 0.2 mL of the ethephon (2-chloroethyl-phosphonic acid) solution onto the filter, **at the last minute** adding 0.2 mL of phosphate buffer (pH 7.5), and **sealing the dish immediately**. Ethephon releases ethylene at this higher pH. As an alternative, several slices of apple peel placed in the petri dish may provide adequate ethylene.

Many additional experiments are possible with this simple system. For instance, one could combine some of the treatments listed above, i.e., add two compounds with opposing effects. Another possibility would be adjusting the concentration of ethylene supplied. You are encouraged to try these or other experiments which might interest you.

(5)   Seal the dishes with Parafilm and take the dishes home with you, keeping them in the dark at room temperature. When the abscission zone is weakened, a jarring of the explants will cause abscission. During the next week, once a day, rap each dish on a flat surface to see if the leaflet portions fall off. Make careful observations of the abscission area, as the lanolin may hold the leaflet in place even after abscission has occurred. Record the number of abscised leaflets (out of twelve) at each observation time for the next two weeks in the chart on page 194. Finally, plot the number of abscised leaflets against time on the graphs that follow. Determine the **rate** of abscission (once it has begun) and record it in the chart on page 194.

PART III. Tissue Printing of Bean Leaf Abscission Zones.

If a section of plant tissue is pressed against a nitrocellulose membrane, an imprint (or anatomical print) of the tissue is left on the membrane; this print may be easily seen using a microscope. In addition, since nitrocellulose membranes readily absorb proteins and nucleic acids from plant tissues, the enzymes and RNA present in the cells are transferred to the print as well. By staining with a general stain for protein or using antibodies specific to a particular protein, the researcher may determine the specific location of enzymes within a plant tissue. You will use petioles from bean plants of different ages and make tissue prints of sections cut from the area of the abscission zone. A general stain for protein will be applied to determine if any changes in protein content have occurred during the aging of the abscission zone.

PROCEDURE

(1)   Use forceps to select a nitrocellulose membrane and its protective cover. Do not touch the membrane with your fingers, as the proteins and nucleic acids present on your hands may be transferred to the membrane and interfere with your observations. Remove the protective cover and set it aside; you will use it later to press the tissue into the membrane.

(2)   Choose a bean plant with intact leaves and one which has had the leaf blades removed earlier in the week. Use a double edge razor blade (be careful) to cut a piece of the petiole containing the abscission zone from one of the plants. The abscission zone is located in the pulvinus (swollen area) at the base of the petiole. Cut the piece vertically, then cut it again to produce a thin vertical section. Use forceps to transfer the section to the piece of nitrocellulose (leave space on the membrane for an additional section). Place the protective paper over the tissue and

press firmly with your index finger to push the tissue into the nitrocellulose. Carefully remove the paper and the tissue (with forceps). Repeat this procedure with an aged petiole, making an additional print on the same membrane. Be sure to mark the edge of the membrane with a pencil so you know which print is which.

(3)   Place the membrane on a clean microscope slide and put the slide on the stage of your microscope (light the sample from the side) and locate each of the tissue prints. Adjust the light and focus up and down to capture as much detail as possible. Locate the abscission zone in each of your tissue prints.

(4)   Place the membrane in a dish of 12 mM HCl (which gives the membrane a negative charge and proteins a positive charge). Pour off the excess acid and, using a transfer pipet, drop several drops of 0.05% CPTS stain on the prints and allow the stain to incubate with the tissue for 2 minutes. Pour off the excess stain, then rinse the membrane several times with 12 mM HCl.

(5)   Again, examine the tissue prints under the microscope. Look for areas of intense blue stain, indicating a high concentration of proteins. Record your observations in the space provided on page 196.

(6)   Repeat all of the steps above, but use a series of transverse sections through the abscission zone.

TEXT REFERENCES

Galston, A. W., P. J. Davies, and R. L. Satter, *The Life of the Green Plant* (3rd ed.), pp. 264-267. Englewood Cliffs, NJ: Prentice Hall, 1980.

Leopold, A. C. and P. E. Kriedemann, *Plant Growth and Development*, pp. 215-220. New York: McGraw-Hill, Inc., 1975.

Salisbury, F. B. and C. W. Ross, *Plant Physiology* (4th ed.), pp. 406-407. Belmont, CA: Wadsworth Publishing Co., 1991.

Taiz, L. and E. Zeiger, *Plant Physiology*, pp. 416, 478-481. Redwood City, CA: Benjamin/Cummings Publishing Co., Inc., 1991.

FURTHER READING

Addicott, F. T., ed., *Abscission*. Berkeley: University of California Press, 1982.

Bickar, D. and P. D. Reid, "A high-affinity protein stain for Western blots, tissue prints, and electrophoretic gels," *Analytical Biochem.* (1992), 203:109-115.

Osborne, D. J., "Abscission," *CRC Critical Reviews in Plant Sciences* (1989), 8:103-129.

EXPERIMENT 22   Name _____

                                                 Date _____

RESULTS

PART I.

Record your data in the chart below and determine the number of days until 50% abscission (i.e., until two of the four petioles have abscised).

Number of Abscised Leaves (of 4)			
Date	Control	Bladeless	Bladeless + NAA
Days to 50% Abscission			

PART II.

Record your data in the chart on the next page. Plot the number of leaflets abscised against time on the graphs that follow. Determine the number of days to 50% abscission (when six out of the twelve leaflets have abscised) and the rate of abscission after abscission has started. NOTE: in some treatments, abscission may not begin for several days; determine the rate of leaflet drop from the graph (as leaflets/day) **after** abscission begins.

Number of Abscised Leaflets (of 12)					
Date	Control	+ NAA	+ Ethephon	+ CHI	+ CHI + Ethephon
Days to 50% Abscission					
Rate of Abscission (leaflets/day)					

Plot the control, the + NAA, and the + CHI treatments on the graph below.

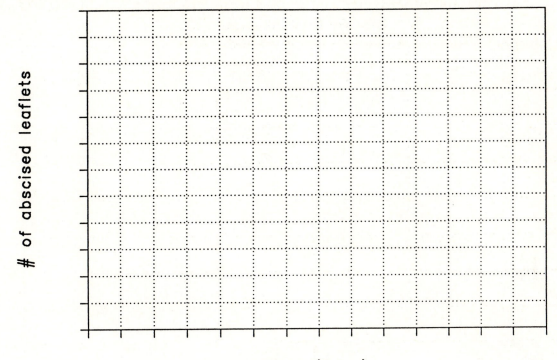

Time (days)

Plot the control, the + ethephon, and the ethephon + CHI treatments below.

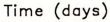

Time (days)

PART III.
Record your observations of unstained and stained tissue prints of bean abscission zones below.

QUESTIONS

1.   In Part I, do you think that the effect of NAA would differ if applied to bladeless petioles of different ages? Why?

2.   In the case of applied auxin, were your results similar for Parts I and II? Why might your results differ?

3.   Can you conclude from your experimental results that auxin is synthesized by the leaf blade and transported to the petiole during the life of the plant? If not, what can be concluded from your results in Part I?

4.   Is abscission an active process rather than one involving merely passive degradation? What evidence from your experiment can you cite in support of your choice?

5.   How would low temperature, respiratory inhibitors, uncouplers, or sulfhydryl reagents (protein denaturation reagents) affect abscission? What biochemical processes would be interfered with in bringing about this effect?

6.   Where was the protein localized in the tissue prints of bean abscission zones? Can you give an explanation of your results?

EXPERIMENT 23

Ethylene Production and Flower Senescence

INTRODUCTION

Ethylene is the most paradoxical of the plant hormones in that, although chemically simple, it displays an extremely wide range of physiological effects, including fruit ripening, leaf and flower senescence, abscission, adventitious root formation, and seed germination. Unlike its counterparts, ethylene is a gas at room temperature. This particular characteristic makes the identification of ethylene a simple matter with the use of gas chromatography.

Gas chromatography is a chemical separation tool which relies on the different affinities that various compounds (in gas form) have for a particular powder (called the "solid" or "stationary" phase). This powder is packed into a six foot long tube (called a "column") and the compounds to be separated are forced through the column under nitrogen gas pressure (N_2 is called the "carrier gas"). The injector and column are usually kept heated. The characteristics of the solid phase determine how fast each gas will pass through, so you will use a solid phase (alumina) which tends to retard ethylene. The constituents of air will exit very quickly while the ethylene will take almost a minute. At the outlet end of the column is a detector; a common one is the flame ionization detector (FID). The FID employs a small hydrogen flame with an anode and a cathode on either side of it. When organic matter is burned in the flame, electrons and ions are released and collect on the anode, producing an electrical current in proportion to the amount of material burned. This current is changed to a voltage and can be recorded on a chart recorder (Fig. 23-1). The nitrogen carrier gas doesn't burn so the recorder will show a flat baseline, but when burnable material goes through the detector it will show up on the recorder as a peak. The area of the peak is proportional to the amount of material burned.

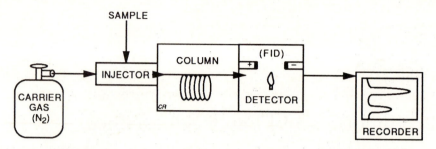

FIGURE 23-1. Schematic illustration of the gas chromatograph (GC).

You will use the gas chromatograph (GC) to measure the production of ethylene in flowers undergoing senescence. Morning Glory (*Ipomoea tricolor*) flowers open at about 6:00 a.m. and begin to close and change color at around 1:00 p.m. Ethylene is rapidly synthesized in the senescing tissue (specifically, the flower ribs) during the afternoon. The pathway for ethylene synthesis is illustrated in Fig. 23-2. Note that the immediate precursor for ethylene is aminocyclopropane carboxylic acid (ACC) and that its synthesis (catalyzed by ACC synthase) is the limiting step in ethylene synthesis. Cytokinins (such as benzylaminopurine) are known to inhibit ethylene synthesis in this system.

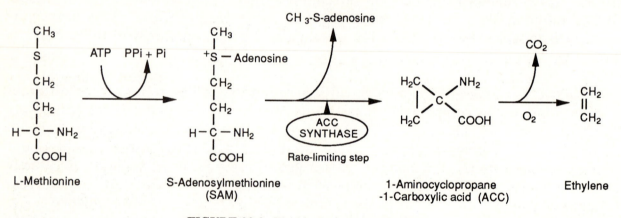

FIGURE 23-2. The biosynthesis of ethylene.

MATERIALS

Equipment
10-mL shell vials
serum stoppers to fit vials
1-mL plastic syringes and 23-G needles*
gas chromatograph

Solutions
5 mM KCl
1 mM 6-benzylaminopurine (BA) in 5 mM KCl
10 μM 1-aminocyclopropane carboxylic
    acid (ACC) in 5 mM KCl

Plant material
Morning Glory (*Ipomoea tricolor* var. Heavenly Blue) flowers

---

*NOTE: The distribution of syringes and hypodermic needles is controlled by law in most states. Be sure to follow your instructor's advice about the storage and disposal of syringes.

PROCEDURE

The Morning Glory flowers you will use started blooming this morning. At 10:00 a.m. they were harvested, and the white rib sections were removed and cut into uniform sections (the blue petal parts have been shown to produce very little ethylene--the rib sections produce most of it). These sections were then floated on a 5 mM KCl solution with or without benzylaminopurine (a cytokinin which inhibits ethylene synthesis in this system) or aminocyclopropane carboxylic acid (the ethylene precursor) and left uncovered for the next three hours (Fig. 23-3). During the first few hours after sectioning, Morning Glory petals produce copious amounts of wound ethylene; the incubation period allows the wound ethylene to dissipate.

(1)   Place five of the rib sections from one of the treatments listed below in a shell vial. Be gentle; further wounding the tissue will increase the amount of ethylene produced regardless of the pretreatment. Position the ribs so that they will not be disturbed by the syringe needle during sampling. Install a serum cap on the vial.

1.   **Control**. The rib sections were floated on 5 mM KCl.

2.   **+BA**. The rib sections were floated on 5 mM KCl with 1 mM BA.

3.   **+ACC**. The rib sections were floated on 5 mM KCl with 10 μM ACC.

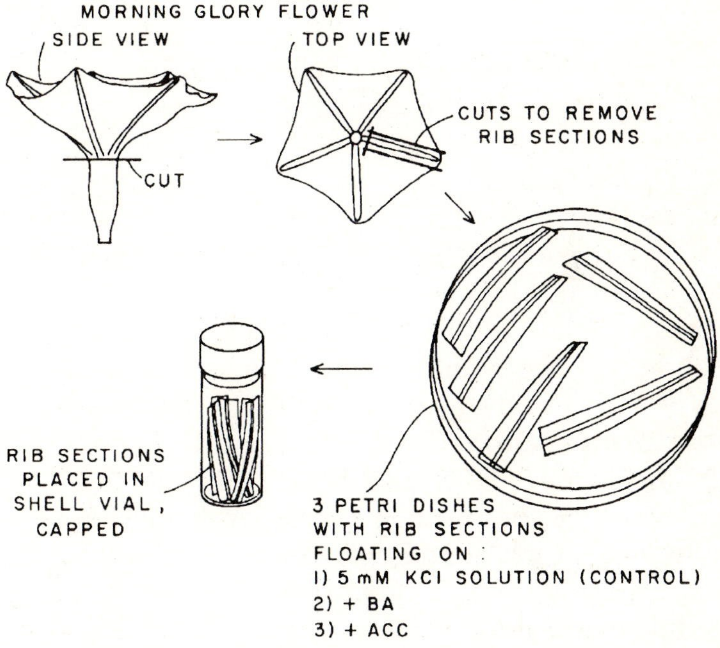

FIGURE 23-3. Preparation of the flower ribs.

(2)   Take samples of the vial atmosphere after one, two and three hours to monitor the progress of the production of ethylene. Be sure to notice the **appearance** (i.e., color and curling) of the petals at each sampling time.

Sampling will be accomplished by withdrawing 1 mL of atmosphere from the vials with a syringe. The syringe must be empty when inserted into the vial, then flushed three or four times with vial atmosphere. The syringe is then filled with 1 mL of vial atmosphere and withdrawn. If necessary, the sample may be stored by gently pushing the tip of the needle into a rubber stopper. Since the vial has a constant volume (9 mL with the septum inserted), removal of 1 mL of atmosphere does two things:

a)   It produces a negative pressure in the vial. After several aliquots have been removed, the septum may begin to leak. To prevent erratic readings resulting from leakage, 1 mL of room air should be injected **into** the vial after each aliquot is removed.

b)   It permanently removes part of the sample. A mathematical correction must be made to account for the lost ethylene (see the Calculations section).

(3)   Your instructor will show you how to operate the GC. The most important aspect of GC sample reproducibility is to insert the syringe, push the plunger, and remove the syringe all in a smooth motion. Another important thing to remember is that the column is under some pressure so a finger must be kept on the end of the plunger to keep it from popping out. The moment the plunger is pressed the chart recorder should be started and the event recorded.

CALCULATIONS

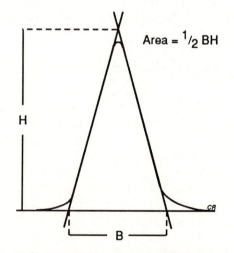

FIGURE 23-4. The area of a triangle.

Modern GCs have peak area integrators (dedicated mini-computers) as a part of their recorders. These recorders instantly print out peak height, area, and retention times. As you may not be so advanced (or rich) you must do these things yourself. The most important parameter to determine is peak area (the retention time for ethylene is only important for identifying which peak is ethylene). The total peak area (Fig. 23-4) is proportional to the total amount of ethylene in the sample. There are two ways to determine peak area manually: 1) cut out the peak and weigh the paper, or 2) calculate the approximate area ($A = \frac{1}{2}BH$). Your instructor will tell you which method to use. A peak generated by 1 mL of a standard mixture of ethylene in air (for example, 1 part per million or ppm which is equivalent to $1\,\mu L/L$) will be posted; this standard can be used to determine the retention time and to calculate the ethylene produced in each treatment.

Since you remove a 1-mL aliquot out of the 9-mL shell vial each time you take a sample, you decrease the concentration of the ethylene remaining in the vial. This ethylene is then missing from all of our subsequent samples. To correct for this destruction of the original sample you must mathematically add back 1/9 (or 0.11) of the area that represents each aliquot removed. Since it is removed for good, it must be added back to **all** subsequent samples. Use the chart on page 203 to record your data and make the corrections at each time point.

If the standard was 1 ppm (or 1 nL/mL) and produced a peak of a given number of chart units (or weight), determine the nanoliters of ethylene present in the 1-mL sample at each sampling time using the following equation:

ethylene per sample (nL/mL) =

$$(1 \text{ nL/mL})(\text{corrected area/standard area}) \qquad \text{Eq. (23-1)}$$

To determine the volume of ethylene produced (in nL) by each rib, multiply the volume of ethylene in each 1-mL sample by the volume of the shell vial (9 mL) and divide by the number of ribs (5):

$$\text{ethylene (nL/rib)} = (\text{ethylene as nL/mL})(9 \text{ mL/5 ribs}) \qquad \text{Eq. (23-2)}$$

The volume of ethylene/rib may be plotted against time (use the graph on page 203) and the **rate** of ethylene production (nL/rib-h) determined.

TEXT REFERENCES

Galston, A. W., P. J. Davies, and R. L. Satter, *The Life of the Green Plant* (3rd ed.), pp. 259-263. Englewood Cliffs, NJ: Prentice Hall, 1980.
Salisbury, F. B. and C. W. Ross, *Plant Physiology* (4th ed.), pp. 393-399. Belmont, CA: Wadsworth Publishing Co., 1991.
Taiz, L. and E. Zeiger, *Plant Physiology*, pp. 473-478. Redwood City, CA: Benjamin/Cummings Publishing Co., Inc., 1991.

FURTHER READING

Kende, H. "Enzymes of ethylene synthesis," *Plant Physiol.* (1989), 91:1-4.
Kende, H. and A. D. Hanson. "Relationship between ethylene evolution and senescence in morning glory flower tissue," *Plant Physiol.* (1976), 57:523-527.
Mattoo, A. K. and J. C. Suttle, eds., *The Plant Hormone Ethylene.* Boca Raton, FL: CRC Press, 1990.
Yang, S. F., and N. E. Hoffman, "Ethylene biosynthesis and its regulation in higher plants," *Annu. Rev. Plant Physiol.* (1984), 35:155-189.

EXPERIMENT 23

Name _____

Date _____

RESULTS

Record the apparent area, then determine the corrected area. Calculate the nL of ethylene in each sample and the volume of ethylene produced per rib at each sampling time.

Treatment _____   Standard area _____

Time	Appearance/ Color	Apparent Area	+ Correction Factors	Corrected Area	Ethylene (nL/mL)	Ethylene (nL/rib)
0		0		0		
1 h		x 0.11				
2 h		x 0.11	+			
3 h			+ +			

Plot your results on the graph below. Assume there was no ethylene in the vial at time zero. Determine the rate of ethylene production per hour.

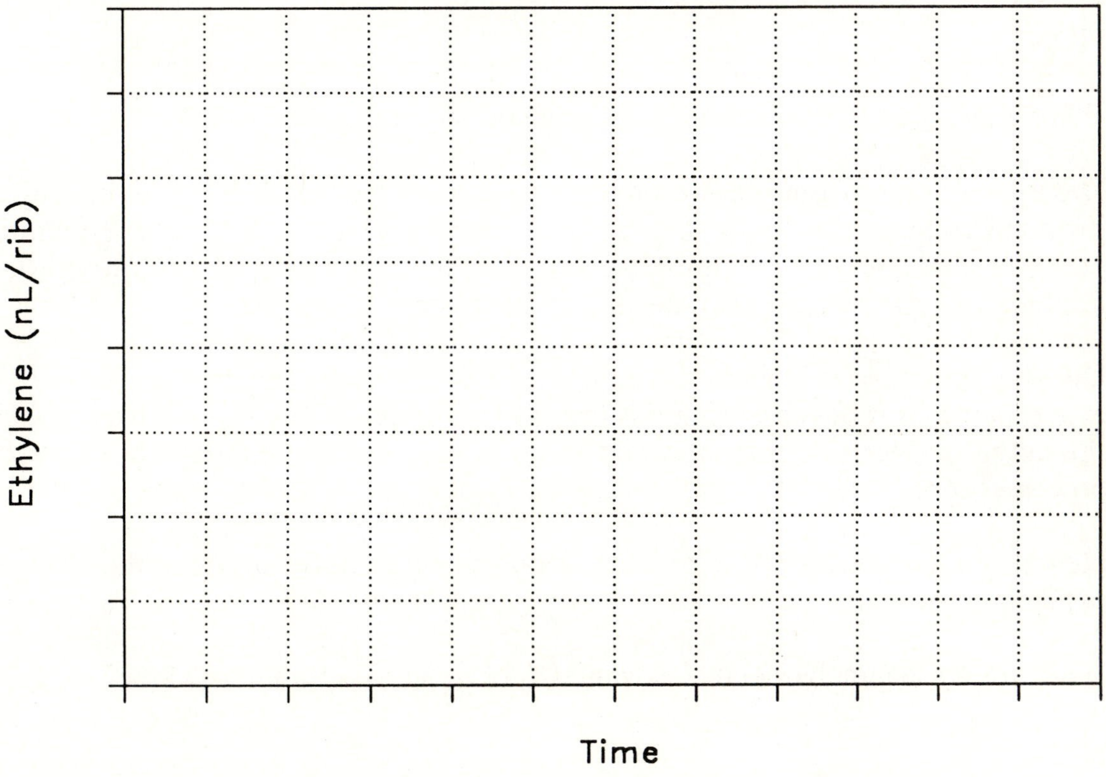

Ethylene (nL/rib)

Time

Show your calculations for your rate determination and record the rate of ethylene production in nL/rib-h. Include the rates for other treatments and observations of the appearance of the ribs for all treatments (use class data) in the chart below.

Rate of ethylene production for your treatment: _____

#	Treatment	Rate (nL/rib-h)	Appearance and Color of Ribs
1	Control		
2	+ BA		
3	+ ACC		

QUESTIONS

1.   What effect did the ACC have on the production of ethylene? Why?

2.   Describe an experiment that would enable you to determine if BA inhibits ethylene production by inhibiting a step before the synthesis of ACC or if BA inhibits ethylene production by inhibiting the last step, the conversion of ACC to ethylene. Explain the logic behind your selection of treatments.

3.   Did you notice any change in the appearance of the petals during the course of the experiment? Is it possible from this experiment to determine if the ethylene acts as the actual trigger of aging, or whether its production is a consequence of earlier processes of senescence?

4.   How might the breakdown of cellular compartmentation (resulting in the change in petal color) enhance senescence?

5.   What are the advantages and disadvantages of using gas chromatography?

EXPERIMENT 24

Bioassay for Gibberellins

INTRODUCTION

One of the pronounced effects of gibberellins (also called gibberellic acid or GA) is the promotion of stem elongation; gibberellins stimulate both cell division and cell elongation. When applied to dwarf plants, gibberellins will induce stem elongation and the plants will become tall. In many long-day plants, applied gibberellins stimulate stem elongation or "bolting."

A biological response, like stem elongation, may be used to measure the amount of a chemical, like gibberellic acid, present in a plant extract. A standard curve for the response may be generated by measuring the extent of hypocotyl elongation of lettuce seedlings grown in solutions containing known amounts of gibberellins. The hypocotyl elongation in a seedling exposed to an **unknown** solution (e.g., a plant extract) may then be compared with the standard curve. This method of estimating the endogenous concentration of a compound, by comparing a biological response with the response of individuals exposed to known amounts of a compound, is called a bioassay.

You will perform a bioassay by growing lettuce seedlings in a series of solutions of known gibberellin concentration and in an unknown solution. When bioassays are used in research, the unknowns are made from plant extracts which are purified (to prevent inhibition of growth by inhibitors and harmful compounds, such as might be found in the vacuole). Purification of plant extracts is time consuming and cannot be performed in the few hours available in the student laboratory; the supplied unknowns for this experiment will be pure compounds obtained from a chemical supply company. The amount of gibberellin will be expressed as gram equivalents of activity, since you are not measuring the actual amount of gibberellins present in your unknown.

MATERIALS

Equipment
24-mL shell vials
2.3-cm filter paper
automatic pipets and tips
clear plastic trays with lids
growth chamber at 23°C, with
fluorescent and gro-lux lamps
rulers

Solutions
gibberellic acid (GA) solutions (see step 2)
unknown GA solution(s)

Plant material
lettuce (*Lactuca sativa* var.
Buttercrunch) seed, germinated 24-26 h.

PROCEDURE

Germinated lettuce seeds will be sown in a series of solutions of known GA concentrations and in an unknown solution.

(1)   Place a small disc of filter paper on the bottom of each of six small vials.

(2)   Label the six vials for the following concentrations of gibberellic acid:

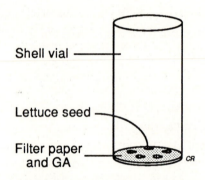

FIGURE 24-1. Shell vial containing five seeds germinating on filter paper.

1.	0	g GA/0.2 mL
2.	10^{-9}	g GA/0.2 mL
3.	10^{-8}	g GA/0.2 mL
4.	10^{-7}	g GA/0.2 mL
5.	10^{-6}	g GA/0.2 mL
6.	unknown	

(3)   Pipet 0.2 mL of the appropriate solutions into the shell vials (Fig. 24-1).

(4)   Place five germinated lettuce seeds on the filter paper in each vial. Choose seeds that have germinated, but whose radicles are less than 2-3 mm long.

(5)   Place the vials in a tray with a transparent cover; the tray will be placed under fluorescent lights (including one "gro-lux" lamp) for four days. After one and three days the lab staff will add 0.1 mL of water to each vial.

(6)   After **four days**, return and measure the length of each hypocotyl (from the root/hypocotyl junction to the cotyledons) to the nearest mm and record your data in the chart on page 209.

CALCULATIONS

Determine the average hypocotyl length for each treatment. Plot the average hypocotyl length for treatments 1 to 5 against the amount of GA (in g/0.2 mL) on the graph on page 210. The amount of GA should be placed on the x-axis on a log scale.

TEXT REFERENCES

Galston, A. W., P. J. Davies, and R. L. Satter, *The Life of the Green Plant* (3rd ed.), pp. 239-241. Englewood Cliffs, NJ: Prentice Hall, 1980.

Salisbury, F. B. and C. W. Ross, *Plant Physiology* (4th Ed.), pp. 375-376. Belmont, CA: Wadsworth Publishing Co., 1991.

Taiz, L. and E. Zeiger, *Plant Physiology*, pp. 433-441. Redwood City, CA: Benjamin/Cummings Publishing Co., Inc., 1991.

FURTHER READING

Takahashi, N., B. O. Phinney and J. MacMillan, eds., *Gibberellins*. New York: Springer Verlag, 1991.

Taylor, A. and D. J. Cosgrove, "Gibberellic acid stimulation of cucumber hypocotyl elongation. Effects on growth, turgor, osmotic pressure, and cell wall properties," *Plant Physiol.* (1989), 90:1335-1340.

Yopp, J. H., L. H. Aung, and G. L. Steffens, eds., *Bioassays and Other Special Techniques for Plant Hormones and Plant Growth Regulators*. Beltsville, MD: Plant Growth Regulator Society of America, 1986.

EXPERIMENT 24   Name _____

                                       Date _____

RESULTS

Record the hypocotyl lengths on the chart below and determine the average hypocotyl length for each treatment. Notice how active gibberellins are and how much growth 1/1,000,000,000 g will produce.

GA (g/0.2 mL)	Hypocotyl Lengths (mm)	Average Length (mm)
0		
10^{-9}		
10^{-8}		
10^{-7}		
10^{-6}		
Unknown ___		

Plot the average hypocotyl length for the first five treatments against the amount of GA in g/0.2 mL on the graph on the next page (note that you are using a log scale). From your standard curve, determine the gibberellic acid concentration in gram equivalents of activity in your unknown. Mark the growth of the unknown on the graph.

Determined concentration of GA in the unknown = _____

Actual concentration of GA in the unknown = _____

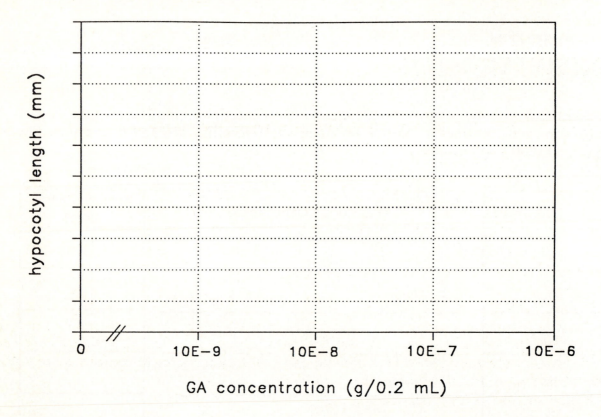

QUESTIONS

1.   Your unknown was prepared from compounds obtained from a chemical supply company. What factors might affect the accuracy of a bioassay if the unknown was a simple plant extract?

2.   Name some advantages and disadvantages of a bioassay compared with the direct measurement of plant hormones by gas chromatography.

EXPERIMENT 25

α-Amylase: Location and Timing in Wheat Seed Germination

INTRODUCTION

Hormones may be defined as chemicals synthesized in one location and translocated to another, where they act in specific ways and at very low concentrations to regulate growth, development and metabolism. One important way by which hormones may act is by stimulating the *de novo* synthesis of specific enzymes. A good example of this process in plants is the stimulation of α-amylase activity in wheat seeds by gibberellic acid (GA).

If you were to measure the level of starch in wheat seeds, you would find that it is high in the ungerminated seed and somewhat lower after several days of germination. The fastest rate of disappearance of starch is found when the embryo is growing and requires the sugars that are produced when starch is broken down. As you might suspect, the activity of the enzyme α-amylase, a digestive enzyme that hydrolyses starch, is correlated with this loss of starch in the endosperm (Fig. 25-1).

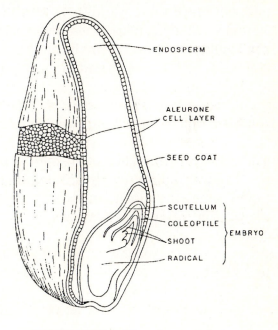

FIGURE 25-1. The wheat seed.

211

De novo synthesis of α-amylase begins in the aleurone layer of the endosperm after imbibition. If the embryo of the seed is removed prior to imbibition, however, no increase in α-amylase activity occurs, suggesting that a hormone or plant growth substance produced by the embryo is necessary for the synthesis of α-amylase. The plant growth substance gibberellic acid (GA) is the messenger that triggers this process. GA is produced in the embryo and migrates to the aleurone layer (Fig. 25-1), where the synthesis of hydrolases (including α-amylase) is stimulated. The hydrolases are then secreted into the endosperm where they cause the breakdown of reserve foodstuffs, particularly starch.

In this experiment you will test the hypothesis that α-amylase is produced *de novo* in the endosperm of wheat seeds using two approaches. First you will make extracts from seeds and, using polyacrylamide gel electrophoresis, determine if the enzyme is present in your extracts. This method allows for the separation of proteins on the basis of molecular weight. A thin gel, constructed from cross-linked polyacrylamide, can act as a three-dimensional sieve for the proteins. A small portion of the protein extract is mixed with sodium dodecylsulfate (SDS) which denatures the proteins and provides them with a very negative charge. The extract is then added to one end of the gel. When the gel is placed in an electric field, the proteins will begin to move toward the positive charge. Large proteins will move slowly through the pores of the gel, while small proteins will be less restricted and will move more rapidly. Once the proteins have separated, they may be visualized by staining. To determine when in the course of germination α-amylase is produced, you will run samples of extract from seed that has been germinated for different time periods and compare them to a standard of pure α-amylase.

Second, to test if **active** enzyme is present at each stage, you will provide starch in a soluble form and add your enzyme extract. When the reaction has run for the prescribed time, it is stopped by the addition of HCl which lowers the pH (remember that enzymes are sensitive to pH). The undigested starch may be measured spectrophotometrically after staining with iodine. It should be noted that you are **not** measuring the specific activity of the enzyme (the enzyme activity per mg total protein) in this experiment, but more likely the concentration of active enzyme.

The experiments in both Parts II and III are designed to give a preliminary answer to several questions:

When does amylase activity become strongest during the course of germination?

Is enzyme activity localized in the endosperm or in the embryo?

Does GA (produced by the embryo) play a role in stimulating the production of α-amylase?

MATERIALS

PART I

Equipment
mortars and pestles, cold
50-mL centrifuge tubes, cold
refrigerated centrifuge
latex gloves
single edge razor blades

Solutions
10 mM citric acid-sodium
    citrate buffer at pH 5
$10 M^{-6}$ M GA in 10 mM buffer at pCa 5

Plant material
wheat (*Triticum aestivum*) seeds, dry and germinated 24, 48, and 72 h.
wheat (*Triticum aestivum*) seeds, incubated at 4°C for 24 h., cut in half, then germinated at room
    temperature for 48 h. +/- GA

PART II

Equipment
electrophoresis apparatus
1.5 mm polyacrylamide gels
power supply
Eppendorf tubes
automatic pipettors
power supply
latex gloves
single edge razor blades
plastic containers or dishes
foam Identi-plugs or white wool
waste container for stain
water bath at 60°C

Solutions
tank buffer: 0.025 M Tris, pH 8.3,
    0.192 M glycine, 0.1% SDS.
2X treatment solution (250 μL aliquots):
    0.125 M Tris-HCl, pH 8.8, 4% SDS,
    20% glycerol, 10% 2-mercaptoethanol,
    pinch Bromophenol blue.
Coomassie Brilliant Blue stain: 0.298%
    Coomassie Blue R, 0.197% Coomassie Blue G,
    20% ethyl alcohol, 7% acetic acid
7% acetic acid (destaining solution)

Cleanup: waste container for Coomassie stain

PART III

Equipment
colorimeters (red filter) and
    tubes or spectrophotometer and tubes
test tubes, 25 x 150 mm
test tube racks
pipets
pipet aids
50- or 100-mL plastic bottles

Solutions
0.05% starch in 0.05 M citric
    acid-sodium citrate buffer, pH 5
1 N HCl
Iodine solution (5 g KI and 0.357 g
    KIO_3 in 1 liter of 2 mM NaOH

PART I. Preparation of the Enzymes. Each group will prepare one or two enzyme extracts (your lab instructor will tell you which ones). All extracts will then be available to each group for Part III of this experiment. Ten different enzyme extracts will be made from wheat seeds as described in Table 25-1. It is essential that the enzyme extract be kept cold during extraction to prevent the action of proteases which might attack the α-amylase. Latex gloves should be worn to protect against proteases as well.

TABLE 25-1. Preparation schedule for enzyme extracts.

#	Seed Part	Germination Time	Special Treatment
1	Whole Seed	0 h	XXX
2	Whole Seed	24 h	XXX
3	Whole Seed	48 h	XXX
4	Whole Seed	72 h	XXX
5	Embryo	72 h	Cut the embryo from the endosperm and grind the embryo only.
6	Endosperm	72 h	Grind the endosperm only.
7	Embryo Half	48 h	24 h at 4°C, then cut in half and germinated.
8	Embryo Half	48 h	24 h at 4°C, then cut in half and germinated in 10^{-6} M GA.
9	Endosperm Half	48 h	24 h at 4°C, then cut in half and germinated.
10	Endosperm Half	48 h	24 h at 4°C, then cut in half and germinated in 10^{-6} M GA.

PROCEDURE

(1)   Wearing latex gloves, select 50 wheat seeds from the container marked for your treatment and place them in a cold mortar. For treatments 5 and 6, select 50 seeds from the 72-hour germinated seeds, use a razor blade to carefully separate the embryo from the endosperm (see Fig. 25-1), and grind each separately. For treatments 7 through 10, take 100 previously cut half seeds. The seeds were soaked for 24 hours at 4 °C; the lower temperature allows the seed to imbibe water without significant germination taking place. The seeds were then separated into embryo and endosperm **halves** and treated as noted (Fig. 25-2). Make sure you select the correct seeds for your enzyme extract.

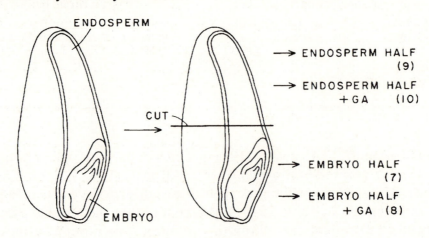

FIGURE 25-2. Seed halves.

(2)   Obtain 40 mL of 10 mM citric acid-sodium citrate buffer at pH 5.0 and add a small portion of the buffer to the seeds. Keeping the mortar cold in an ice bucket, grind the seeds thoroughly, adding more fluid as you go along until about 30 mL have been added.

(3)   Transfer your homogenate to a 50-mL centrifuge tube labeled for your treatment, then use the last 10 mL of buffer to rinse the mortar, adding this rinse to the centrifuge tube too. The instructor will collect all the tubes and sediment them for 10 minutes at 15,000 g in the cold to remove starch grains, cell walls, mitochondria and nuclei. While the extracts are in the centrifuge, begin the set up for Parts II and III and read the appendix on the use of pipets (Appendix B).

(4)   The supernatant from each tube will be poured into a labeled container and placed in an ice bucket for everyone to take a sample. A labeled pipet will be placed in each. Be sure to use only the appropriate pipet for each homogenate.

PART II. The Separation of Proteins by Polyacrylamide Gel Electrophoresis (PAGE).
This part of the experiment is designed to determine if α-amylase is present in the extracts you made in Part I.

PROCEDURE

(1)   Add 250 µL of the enzyme extract(s) you prepared to an equal amount of 2X treatment buffer (which contains a marker dye) in an Eppendorf tube.

(2)   The instructor will prepare the electrophoresis apparatus (Fig. 25-3) and demonstrate how to load your sample onto the gel. Use an automatic pipettor and a fresh pipet tip to load 25 µL of the enzyme mixture into the appropriate lane in the stacking gel at the top. A 25-µL sample of an α-amylase standard will be loaded into one lane as well.

(3)   When all of the lanes are filled, the instructor will set the proper

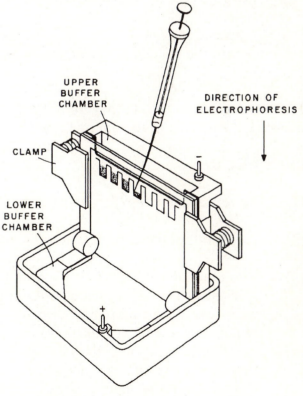

FIGURE 25-3. Electrophoresis apparatus.

current and the gel will run for about 1 hour (continue with Part III of the experiment). When the marker dye has reached the bottom of the gel, the power supply will be disconnected.

Divide the following steps among the class:

(4)   Wearing latex gloves, carefully remove the glass plate from the top of the gel, by prying gently at the bottom edge. Cut off the stacking gel pieces that remain and notch the upper left hand corner of the gel to identify the orientation of the gel.

(5)   Place the gel in a plastic container or dish with enough of the stain solution to cover the gel; incubate the gel for 10 minutes at room temperature.

(6)   Pour off the stain (into the appropriate waste container), rinse once with distilled water, then cover the gel with destaining solution. Add several foam plugs or pieces of wool to speed up the destaining. Place the plastic container in a 60 °C water bath for about an hour. As the stain is removed from the gel, the bands of protein will retain the stain and become visible. Note the position of the α-amylase standard band and look for a band at the same position in the other lanes. Leave the gel in the destaining solution at 60 °C overnight.

Examine the gels on the following day. Determine which bands are α-amylase and record the bands you see on the templates on page 219.

PART III. An Assay of Enzyme Activity. This part of the experiment is designed to measure the **activity** of the enzyme extracts you made in Part I.

PROCEDURE

(1)   Label 12 test tubes (25 x 150 mm), following the protocol listed in Table 25-2.

(2)   Add 1 mL of the appropriate enzyme extract to each test tube.

(3)   The reaction will start when you add 2 mL of starch (0.05% "soluble starch" in 0.05 M citric acid-sodium citrate buffer at pH 5.0) to the enzyme solution. The instructor will have determined the appropriate reaction time for your enzymes. If the reaction time is long (several minutes), reasonably accurate timing can be obtained by adding starch in a timed sequence to the 12 tubes already containing enzyme; then at the end of the reaction time going down the line of tubes adding HCl (step 4) in the **same** sequence. If the enzyme is very active and the reaction time short, each reaction may be performed and timed individually. Note that the reaction time varies for different treatments.

TABLE 25-2. Protocol for the essay for enzyme activity, Part III.

#	Hours Germinated	Part Used	Enzyme Extract	Starch	Reaction Time*
1	0	Whole Seed	1 mL	2 mL	1 min
2	24	Whole Seed	1	2	1
3	48	Whole Seed	1	2	1
4	72	Whole Seed	1	2	1
5	72	Embryo	1	2	1
6	72	Endosperm	1	2	1
7	48	Embryo Half	1	2	5
8	48	Embryo Half + GA	1	2	5
9	48	Endosperm Half	1	2	5
10	48	Endosperm Half + GA	1	2	5
11	72	Whole Seed	1	2	0**
12	72	Whole Seed	1	0***	1

*   Suggested reaction times. Your instructor will inform you of the appropriate reaction times for your class.
**   Add HCl before adding the enzyme extract.
*** Add 2 mL of H_2O instead.

(4)   At the end of the reaction time, add 7 mL of 1 N HCl to the tubes (except 11 which received the HCl earlier) and **mix**.

(5)   Add 1 mL of iodine solution to the killed reaction mixture in each test tube. Iodine develops a blue color when mixed with starch; if the starch has been hydrolyzed by enzyme activity, less blue color will be produced.

(6)   Transfer the reaction mixes to a colorimeter or spectrophotometer tube or cuvette. Measure the absorbance of the blue solution, using a red filter in the colorimeter or a spectrophotometer set at 580 nm. Record your results in the table on page 220. If the solutions are too dense to measure readily, dilute them by a measured amount.

CALCULATIONS

Correct for background by subtracting the value for tube 12 from the values for all other tubes.

Assume that the colorimeter units (or absorbances) you measured are all due to the starch in the substrate solution added, and that there was none lost in the 0-time control (tube 11). With the value for tube 11 as 100%, calculate the % of starch disappearing in each case.

If X represents the amount of starch present at the beginning of the experiment (tube 11) and Y represents the amount of starch remaining at the end of the experiment (separately in tubes 1-10) then:

$X - Y = Z$, where Z represents the **amount** of starch lost during the experiment

and

the % of starch lost = $100 (Z/X)$

If more than 90% of the starch disappeared with any one enzyme preparation, consider the results to be enzyme saturated, thus comparing results of more than 90% to each other may be invalid since the reaction may have been limited by substrate.

Plot the determined enzyme activity (% starch lost) as a bar graph on the graphs on page 221.

TEXT REFERENCES

Galston, A. W., P. J. Davies, and R. L. Satter, *The Life of the Green Plant* (3rd ed.), pp. 244-247. Englewood Cliffs, NJ: Prentice Hall, 1980.

Salisbury, F. B. and C. W. Ross, *Plant Physiology* (4th ed.), pp. 376-380. Belmont, CA: Wadsworth Publishing Co., 1991.

Taiz, L. and E. Zeiger, *Plant Physiology*, pp. 444-448. Redwood City, CA: Benjamin/Cummings Publishing Co., Inc., 1991.

FURTHER READING

Atzorn, R. and E. W. Weiler, "The role of endogenous gibberellins in the formation of α-amylase by aleurone layers of germinating barley caryopses," *Planta* (1983), 159:289-299.

Chrispeels, M. J. and J. E. Varner, "Gibberellic acid-enhanced synthesis and release of α-amylase and ribonuclease by isolated barley aleurone layers," *Plant Physiol.* (1967), 42:398-406.

Fincher, G. B., "Molecular and cellular biology associated with endosperm mobilization in germinating cereal grains," *Annu. Rev. Plant Physiol. Plant Mol. Bio.* (1989), 40:305-346.

EXPERIMENT 25

Name _____
Date _____

RESULTS

PART II.

Use the templates below to record the positions of the α-amylase bands on the gels. Note which extracts contain the enzyme and which do not.

| Std | 0 h | 24 h | 48 h | 72 h | Emb | Endo |

| Std | Emb Half | Emb Half +GA | Endo Half | Endo Half +GA |

PART III.

Record the colorimeter units (or A) and determine the % starch lost for each treatment.

#	Enzyme extract	Colorimeter Units or A_{580}	– Background = Y	X – Y = Z	% Starch Lost (100 Z/X)
1	0 h				
2	24 h				
3	48 h				
4	72 h				
5	Embryo				
6	Endosperm				
7	Embryo Half				
8	Embryo Half + GA				
9	Endosperm Half				
10	Endosp Half + GA				
11	0-Time Control		= X		
12	Background				

Plot the % starch lost for each treatment as a bar graph on the next page.

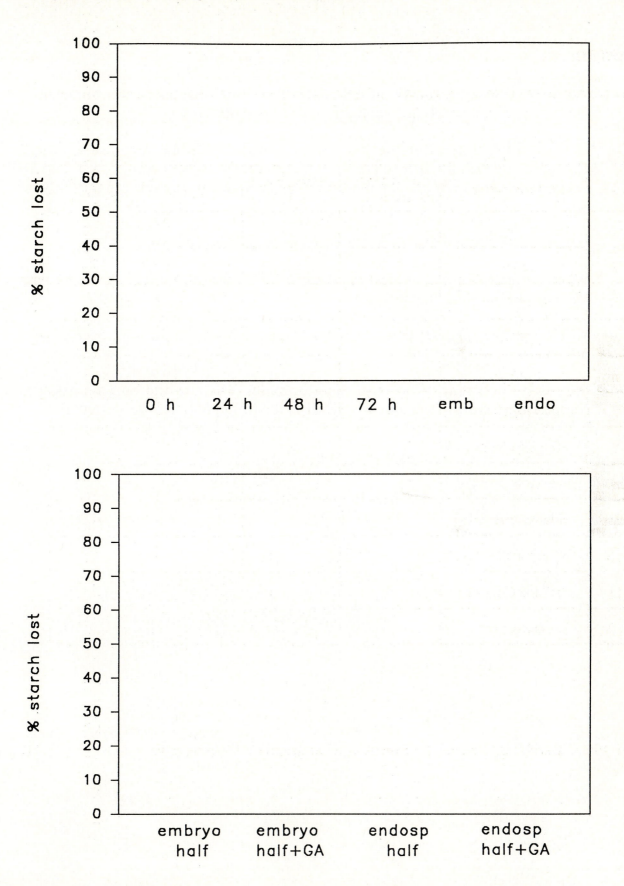

QUESTIONS

1. When does amylase activity become strongest during the course of germination? Do your results support the hypothesis of *de novo* synthesis?

2. Is there a lag time after the imbibition of water before α-amylase appears? Compare your results from Part II (electrophoresis) to your results from Part III (enzyme activity). What explanation can you give for any discrepancy in your results?

3. Is there more enzyme activity in the endosperm or in the embryo?

4. What is the evidence to indicate that some signal must move into the endosperm half before α-amylase can appear?

5. What evidence suggests that GA is that signal? Where in the seed does the GA originate?

6. Why is it important to do both the electrophoresis and the enzyme assay to characterize the involvement of α-amylase in the germination of wheat seeds?

EXPERIMENT 26

Phytochrome Control of Leaflet Movement in *Albizzia*

INTRODUCTION

The leaflets of *Albizzia* exhibit diurnal "sleep movements"; the leaflet pairs close together at night and open during the day. These movements are partially controlled by a circadian rhythm. If plants which have been in the greenhouse are placed in constant darkness, they will continue to open and close on a 24-hour cycle, although the response will eventually fade out. However, since the light regime "sets" the biological clock, a response to changing light and dark may be observed. The pigment phytochrome (Fig. 26-1) has been implicated as the effective pigment involved in this phenomenon. Photoreversibility (i.e. a dependence upon the last light received **only**) is considered good evidence for the involvement of phytochrome in any system.

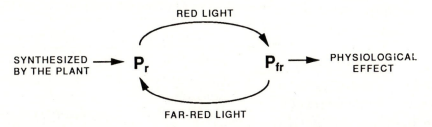

FIGURE 26-1. The oldest and simplest model for the photoreversibility of phytochrome

The movement of leaflets in *Albizzia* is a result of differential turgor changes in motor cells situated on the upper and lower sides of the pulvinule at the base of each leaflet. This differential turgor change has been shown to be produced by the movement of K^+ ions into or out of the cells under the influence of light or endogenous rhythm. The rate and extent of leaflet closure is influenced by the percent of the phytochrome in the P_{fr} form when the plants are placed in the dark. Since the P_{fr} level is high following the red irradiation and low following far-red irradiation (Fig. 26-1), leaflet movement can be studied by irradiating leaflets with red or far-red light, then measuring the degree of closure during a subsequent dark period.

The phytochrome effect is most apparent when the treatments are given early in the light period (otherwise the endogenous rhythm effect predominates), so you will be given plants which were illuminated until one or two hours before class time in the growth chamber for about a week prior to the experiment. Plants should remain in the chamber until needed and prepared as rapidly as possible in as bright a light as possible, or the leaflets may close.

MATERIALS

Equipment
growth chamber
plastic petri dishes
razor blades
red-light source
far-red-light source
darkroom with green safelight
light-tight cabinet or box

Plant material
Albizzia julibrissin plants
maintained on a 12-hour
photoperiod with the light
period beginning one or two hours
before class

PROCEDURE

(1)   Label six petri dishes for the treatments listed in Table 26-1.

(2)   Remove two pairs of pinnae from open leaves of intermediate age of an *Albizzia* plant. Discard the terminal pair and the two basal pinnule pairs,

TABLE 26-1. Protocol for light treatments.

#	Treatment	Light Regime
1	Red	5 min Red
2	Red, Far-Red	5 min Red, 5 min Far-Red
3	Far-Red	5 min Far-Red
4	Far-Red, Red	5 min Far-Red, 5 min Red
5	Dark	Place in the dark
6	Light	Keep on the lab bench in the light

then segment the remaining leaflets into pinnule pairs, each attached to a small piece of rachilla (Fig. 26-2). Select pinnule pairs randomly and place them on water in the six dishes (six pairs per dish). Work rapidly to prevent the leaflet pairs from drying out.

(3)   Take dishes 1 to 5 to the dark room and leave dish 6 in the light in the lab. Place dish 5 in a light-tight cabinet and treat the others according to the schedule above.

Be sure that no leaflets receive white light **after** they have been treated with the appropriate light treatment.

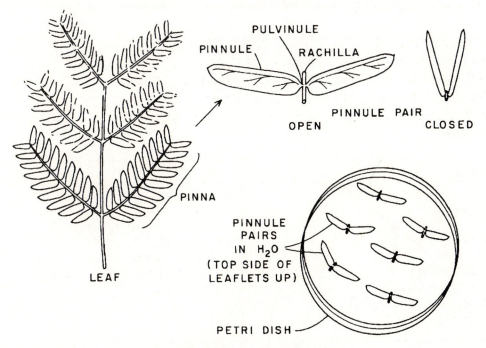

FIGURE 26-2. Pinnule pairs.

(4) Place all the dishes in the dark and inspect the red-light-treated pinnules at 10-minute intervals using only the green safelight. When the red-light-treated pinnules have closed to about a 60° angle between the two pinnules of a pair (after 10 to 60 minutes), take all the pinnule pairs to the classroom.

(5) Rapidly record the angles between the leaflets. Handle the leaflets with forceps and hold them next to the protractor in Fig. 26-3 to determine the angle. Read one from each petri dish in turn (to account for continued closing during measurement) until all the angles have been measured. Record your data in the chart on page 227.

TEXT REFERENCES

Galston, A. W., P. J. Davies, and R. L. Satter, *The Life of the Green Plant* (3rd ed.), pp. 300-318. Englewood Cliffs, NJ: Prentice Hall, 1980.

Salisbury, F. B. and C. W. Ross, *Plant Physiology* (4th ed.), pp. 408-411. Belmont, CA: Wadsworth Publishing Co., 1991.

Taiz, L. and E. Zeiger, *Plant Physiology*, pp. 497-505. Redwood City, CA: Benjamin/Cummings Publishing Co., Inc., 1991.

FURTHER READING

Bioscience (1983), 33. The entire issue is devoted to the discussion of biological clocks.

Kim, H. Y., G. G. Coté and R. C. Crain, "Potassium channels in *Samanea saman* protoplasts controlled by phytochrome and the biological clock," *Science* (1993), 260:960-962.

Lumsden, P. J. "Circadian rhythms and phytochrome," *Annu. Rev. Plant Physiol. Plant Mol. Bio.* (1991), 42:351-371.

Satter, R. L. and A. W. Galston, "Mechanisms of control of leaf movements," *Annu. Rev. Plant Physiol.* (1981), 32:83-110.

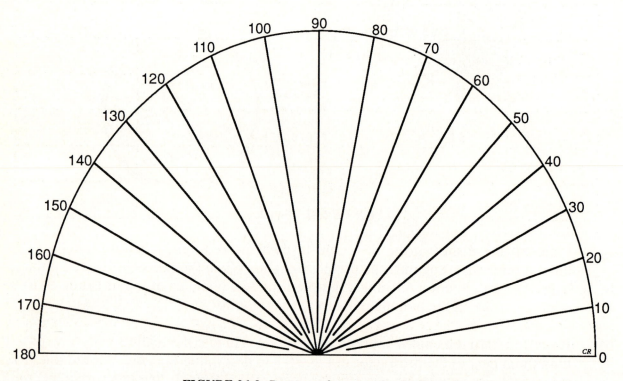

FIGURE 26-3. Protractor for measuring leaflet angles.

EXPERIMENT 26

Name _____

Date _____

RESULTS

Record the leaflet angles in the chart below and determine the average angle between the leaflets for each treatment.

#	Treatment	Leaflet Angles	Average Leaflet Angle
1	Red		
2	Red, Far-Red		
3	Far-Red		
4	Far-Red, Red		
5	Dark		
6	Light		

QUESTIONS

1. Does the evidence derived from your data indicate a phytochrome involvement in leaflet movement? Briefly explain.

2. How might phytochrome act to effect these movements?

3. What is the sequence of events which result in leaflet closure? i.e., why are pulvinus (or pulvinule) cells often referred to as "motor cells"?

4. What evidence can you cite from discussion or lecture to demonstrate that an endogenous, circadian (i.e., diurnal) rhythm also operates in sleep movements?

EXPERIMENT 27

Seed Germination: Light and Hormones

INTRODUCTION

Dormancy of seeds is defined as the failure to germinate even under optimum conditions of moisture, temperature and aeration. Dormant lettuce seeds (var. Grand Rapids) require light after imbibition before they will germinate; the effective light is red, implicating phytochrome. The mechanism that controls dormancy and the role of light in breaking dormancy remain unclear; however, hormonal involvement is likely as gibberellic acid (GA) has been shown to break dormancy in some light-sensitive seeds even in the dark. The light sensitivity of lettuce seeds depends very much on the prior history of the seeds; if the seeds have been stored too long under warm, moist conditions or if they have been exposed to cold temperatures, internal changes may have occurred which enable them to germinate without light.

The layers surrounding the embryo clearly play a role in maintaining dormancy. Isolated embryos grown in the dark will germinate, but if the embryos are placed in a solution with a negative water potential (e.g., 0.1 or 0.2 M mannitol), germination is again inhibited in the dark. Embryos grown in mannitol in the light germinate. Apparently, exposure to red light increases the capacity of the embryos for growth, possibly by making the water potential more negative and enabling the embryo to take up water from the surrounding solution.

Grand Rapids lettuce seed was used for important work in the characterization of the pigment phytochrome in the 1930s and is an excellent system for showing photoreversibility (see Expt. 26). You will test for photoreversibility and for the effect of several plant growth substances on germination of light-sensitive seed in the dark.

MATERIALS

Equipment
petri dishes, 9 cm
No. 1 filter paper, 9 cm
pipets
Parafilm
red-light source
far-red-light source
light-tight cabinet or box
darkroom

Solutions
600 μM gibberellic acid (GA)
   in 10 mM phosphate buffer
120 μM abscisic acid (ABA) in
   10 mM phosphate buffer
12 μM kinetin (CK) in 0.1 M
   phosphate buffer
(0.2 M mannitol)

Plant material
light sensitive lettuce (*Lactuca sativa* var. Grand Rapids) seed soaked for 3 hours in dark in 9-cm petri
   dishes containing three layers of filter paper
light sensitive lettuce (*Lactuca sativa* var. Grand Rapids) seed, dry

PART I. The Effect of Light on the Germination of Lettuce Seeds var. Grand Rapids.

PROCEDURE

(1)   In the dark room, using only a green safelight, remove six petri dishes, each containing imbibed lettuce seeds (var. Grand Rapids or other light sensitive seed) from a light-tight box. Each dish should contain approximately 50 seeds. Expose the seeds to the light regimens listed in Table 27-1.

Make sure each dish has sufficient water (add more if needed) and seal the dish with Parafilm. After the prescribed light treatments, wrap the petri dishes for treatments 1-5 in foil (they should receive no further light).

(2)   After three days, return to the laboratory and open the petri dishes. Count the total number of seeds and the number of seeds that have germinated. Record your data in the chart on page 233 and determine the percent germination for each treatment.

TABLE 27-1. Protocol for light treatments, Part I.

#	Treatment	Light Regime
1	Red	5 min Red
2	Red, Far-Red	5 min Red, 15 min Far-Red
3	Far-Red	15 min Far-Red
4	Far-Red, Red	15 min Far-Red, 5 min Red
5	Dark	Keep in the dark
6	Light	Place on the lab bench in the light

PART II. The Effects of Inhibitors and Promoters on Germination of Lettuce Seed var. Grand Rapids.

PROCEDURE

(1) Line four petri dishes with three layers of filter paper each. Add the appropriate solutions for the treatments listed in Table 27-2. Additional experiments are possible using this system; if you are interested, make up several more petri dishes and combine several of the treatments.

TABLE 27-2. Protocol for Part II, lettuce seed germination.

#	Treatment	GA	ABA	CK	H$_2$O
1	GA, Dark	2 mL	0	0	4 mL
2	CK, Dark	0	0	2 mL	4 mL
3	ABA, Dark	0	2 mL	0	4 mL
4	ABA, Light	0	2 mL	0	4 mL

(2) Sow 50 dry lettuce seeds in each petri dish. Seal with Parafilm. This step may be completed in the light as the seed is not light-sensitive until it has imbibed water.

(3) Wrap all the dishes (except dish 4) in foil and place them in a dark cabinet; place dish 4 in the light.

(4) Assess germination (as %) and growth after three days. Treatments 5 and 6 of Part I will be your dark and light controls. Record your data in the chart on page 233.

TEXT REFERENCES

Galston, A. W., P. J. Davies, and R. L. Satter, *The Life of the Green Plant* (3rd ed.), pp. 285-292. Englewood Cliffs, NJ: Prentice Hall, 1980.

Salisbury, F. B. and C. W. Ross, *Plant Physiology* (4th ed.), pp. 379-383. Belmont, CA: Wadsworth Publishing Co., 1991.

Taiz, L. and E. Zeiger, *Plant Physiology*, 490-493. Redwood City, CA: Benjamin/Cummings Publishing Co., Inc., 1991.

FURTHER READING

Borthwick, H. A., S. B. Hendricks, M. W. Parker, E. H. Toole and V. K. Toole, "A reversible photoreaction controlling seed germination," *Proc. Nat. Acad. Sci.* (1952), 38:662-666.

Carpita, N. C. and M. W. Nabors, "Growth physics and water relations of red-light-induced germination in lettuce seeds. V. Promotion of elongation in the embryonic axes by gibberellin and phytochrome," *Planta* (1981), 152:131-136.

Smith, H. *Phytochrome and Photomorphogenesis*, Chap. 6. London: McGraw-Hill, 1975.

EXPERIMENT 27

Name _____

Date _____

RESULTS

PART I.

Enter your data into the chart at the
right and determine the % germination
for each treatment.

#	Treatment	# Germinated / # Seeds	% Germination
1	Red		
2	Red, Far-Red		
3	Far-Red		
4	Far-Red, Red		
5	Dark		
6	Light		

PART II.

Record your data in the chart below. Include an assessment of growth and determine the
% germination in each case.

#	Treatment	# Germinated / # Seeds	% Germination	Assessment of Growth (development of radicle)
1	GA, Dark			
2	CK, Dark			
3	ABA, Dark			
4	ABA, Light			

QUESTIONS

1.   Does your data indicate that lettuce seed germination is phytochrome mediated? How can you tell? If so, what form of phytochrome is predominant in causing germination?

2.   What might the ecological significance of this adaptation be to the plant?

3.   Can gibberellic acid substitute for light in the germination of lettuce seeds? Can kinetin? How might GA or kinetin stimulate germination?

4.   Can you conclude from our experiments that phytochrome acts to stimulate hormone production *in situ*, and the hormone then induces germination?

5.   Germination is observed when the primary embryonic root (radicle) grows through the surrounding tissue and seed coat. The endosperm in lettuce seeds is composed of cells with thickened walls which resist growth of the embryo axis. For germination to occur, either the capacity of the axis to penetrate (or physically push its way through) the endosperm must be increased or the capacity of the endosperm to resist embryo growth must be reduced. From what you know of the effects of GA in other systems (or from other experiments in this book), how might GA be capable of causing both alternatives?

EXPERIMENT 28

De-etiolation and Phytochrome

INTRODUCTION

Plants grown completely in the dark do not develop the same morphology as those grown in the light, although the extent of the difference depends somewhat on the species. The plants in the dark are called "etiolated," the phenomenon "etiolation." In general usage this term refers especially to the lack of green color, but also may refer to morphological differences. These plants are tall and slender with a recurved tip (epicotyl or hypocotyl hook) and have small leaves lacking in pigments. Not very much light is needed to reverse the effect of complete darkness (Fig. 28-1); the effective light is red and it is in turn reversed by far-red illumination, hence the pigment phytochrome is presumed to be the photoreceptor (see Expt. 26).

When etiolated plants are exposed to light, changes in morphology occur within a few days. The growth of the stem slows, the hook at the apex "opens" and the leaves expand and become green (see Expt. 29). Changes in several of these characteristics may be used as a measure of the degree of "de-etiolation."

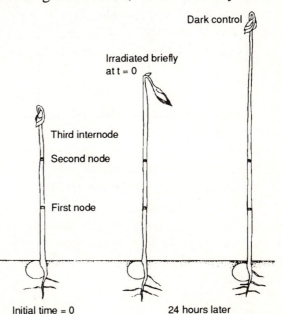

FIGURE 28-1. Etiolated and treated plants. (Adapted from Galston, Davies & Satter, *The Life of the Green Plant*, p. 286, Prentice Hall, 1980.)

235

MATERIALS

Equipment
dark room with green safelight
red light source
far-red light source
pot sticks

Plant material
Red Kidney bean (*Phaseolus vulgaris* var. Redcloud) and pea (*Pisum sativum*) plants grown in the dark

PROCEDURE

Listed below is a protocol of light treatments designed to test for photoreversibility and answer the question "Is phytochrome the effective pigment in the apical hook opening of pea and bean plants?" The question may be answered by giving alternating red and far-red light treatments, then examining the plants after three days for changes in the curvature of the apical hook (See Fig. 28-1 and Appendix F; peas have an epicotyl hook and beans have a hypocotyl hook).

1. Dark control. No irradiation.
2. Light control. Plants placed in daylight.
3. 15 minutes red light
4. 15 minutes red light + 15 minutes far-red light
5. 15 minutes far-red light
6. 15 minutes far-red light + 15 minutes red light

Rather than testing for photoreversibility, assume that phytochrome is the effective pigment in de-etiolation and devise another experiment to give more information about the phytochrome system. That is, ask a question and answer it experimentally. Some possible questions are: How many minutes of red light are required to cause a specific response? Is there an escape time, when far-red light can no longer reverse the effects of red light? Is sunlight or red light more effective at reversing the response? How many minutes of far-red light are required to reverse the effects of red light? Is the effect of repeated red light treatments additive? Answer one of these questions or devise and answer your own question. Choose peas or beans and examine one characteristic three days after your light treatments.

It is suggested that a sign-up sheet for darkroom time be made available and that you make your light exposures and observations sometime before the next laboratory class. Inform your instructor if you need any special equipment such as protractors or rulers.

TEXT REFERENCES

Galston, A. W., P. J. Davies, and R. L. Satter, *The Life of the Green Plant* (3rd ed.), pp. 296-297. Englewood Cliffs, NJ: Prentice Hall, 1980.

Salisbury, F. B. and C. W. Ross, *Plant Physiology* (4th Ed.), pp. 457-459 and 462-463. Belmont, CA: Wadsworth Publishing Co., 1991.

Taiz, L. and E. Zeiger, *Plant Physiology*, pp. 490-494. Redwood City, CA: Benjamin/Cummings Publishing Co., Inc., 1991.

FURTHER READING

Downs, R. J., "Photoreversibility of leaf and hypocotyl elongation of dark-grown Red Kidney bean seedlings," *Plant Physiol.* (1955), 30:468-472.

Downs, R. J., S. B. Hendricks, and H. A. Borthwick, "Photoreversible control of elongation of pinto beans and other plants under normal conditions of growth," *Bot. Gaz.* (1957), 118:199-208.

Furuya, M., ed., *Phytochrome and Photoregulation in Plants.* Tokyo: Academic Press, 1987.

EXPERIMENT 28

Name _____

Date _____

RESULTS

Use the space below to outline your protocol. Put your results in table form, then, if appropriate, graph the response (change in height, for example) against the parameter you have chosen (duration of light exposure, for example).

QUESTIONS

1. What is the ecological significance of phytochrome control of de-etiolation?

2. Explain how your experimental results serve to answer the question you have devised.

3. Considering the information you have acquired with your results, briefly describe a new experiment you might perform to further define how the phytochrome system works in your plants. Assume no limit to the available equipment.

EXPERIMENT 29

The Greening of Cucumber Cotyledons

INTRODUCTION

Chlorophylls are the end product of a biosynthetic pathway which is regulated to a large extent at the first committed biochemical step: the conversion of the amino acid glutamate to δ-aminolevulinic acid (ALA). Both cytokinins (plant hormones) and light stimulate the production of ALA and plastid biogenesis at several points. The observable result is that they will greatly stimulate the greening process.

Plants which are grown completely in the dark lack chlorophyll and are called "etiolated." Etiolated cotyledons will green more rapidly when placed in the light if they have been pretreated with cytokinin. Similarly, a pretreatment with light (specifically red light, absorbed by the pigment phytochrome) will shorten the time to greening. If the cotyledons from seedlings grown in the dark are provided ALA exogenously, greening (in this case, the production of protochlorophyllide) will occur even in the dark. A protein synthesis inhibitor such as cycloheximide (CHI) greatly reduces chlorophyll synthesis.

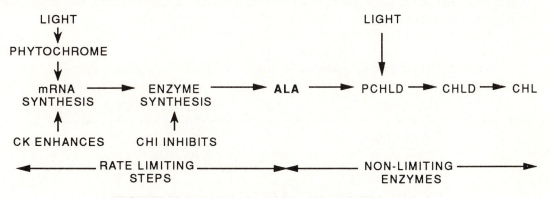

FIGURE 29-1. Proposed model for chlorophyll biosynthesis.

Examine the model for chlorophyll biosynthesis (Fig. 29-1) and read over the procedure. Choose several treatments from the list provided in Table 29-1 (or suggest your own!) and try to support or refute some aspect of the model.

MATERIALS

Equipment
plastic petri dishes, 9 cm
No. 1 filter paper, 9 cm
Parafilm
pipets
darkroom
foil
red-light source
room light
balance
mortars and pestles
funnels
spectrophotometer
test tubes or cuvettes
    for spectrophotometer

Solutions
6-benzylaminopurine (BA), 5 mg/L in 20 mM
    HEPES buffer, pH 7
δ-aminolevulinic acid (ALA), 2 g/L in 20 mM
    HEPES buffer, pH 7
cycloheximide (CHI), 200 mg/L in 20 mM
    HEPES buffer, pH 7
20 mM HEPES buffer, pH 7
90% acetone containing 10 mM
    NH_4OH

Plant material
6 day old cucumber (*Cucumis sativus* var. Marketmore 70)
seedlings, grown in the dark at 26°C

PROCEDURE

(1)   Determine which aspect of the model you want to test and plan your experiment. Select the treatments that are needed to answer your question from Table 29-1 or develop your own treatments. Be sure to include the proper control treatments (it is suggested that **all** groups do treatments 1 and 3). Discuss your experimental plan with your instructor.

TABLE 29-1. Suggested treatments for testing the model for chlorophyll synthesis.

#	Treatment	CK	ALA	CHI	HEPES Buffer
1	Dark Control	0	0	0	6 mL
2	Dark + CK	3 mL	0	0	3 mL
3	Control, 2 h Light	0	0	0	6 mL
4	+ CK + 2 h Light	3 mL	0	0	3 mL
5	2 min Red + 2 h Light	0	0	0	6 mL
6	+ ALA + 2 h Light	0	3 mL	0	3 mL
7	+ CHI + 2 h Light	0	0	3 mL	3 mL
8	+ ALA + CHI + 2 h Light	0	3 mL	3 mL	0

(2)   Prepare the appropriate number of petri dishes, by placing three layers of filter paper in each and labeling for the treatment given. Add the appropriate amounts of solutions for your treatments.

(3)   In the dark room under a green safelight, excise the cotyledon pairs from etiolated cucumber seedlings, breaking the hypocotyl at the point where the cotyledons join. Use cotyledons which are still appressed together and place seven cotyledon pairs in each petri dish. Wrap your stack of petri dishes with foil and leave it overnight in the dark at 26 °C.

(4)   The following morning, place your petri dishes (except for treatments 1 and 2) in room light for two hours. Note any obvious changes in color which may have occurred overnight. To obtain consistency in class results, careful timing of the two hour light period is essential.

(5)   At the end of the two hour light period, blot the cotyledons on filter paper and weigh each group of seven. Record the weights in the chart on page 245.

(6)   Place each set of cotyledons in a mortar, cover with liquid N_2 (if available) and grind the cotyledons to a powder (liquid N_2 makes the process easier but is not required). Work quickly and, if possible, grind each set of cotyledons concurrently.

(7)   Add 3 mL of 90% acetone and grind completely.

(8)   Filter through acetone-premoistened filter paper into a graduated test tube, then bring the total volume to 5 mL.

(9)   Read the absorbance on a spectrophotometer at 652 nm (the wavelength where chlorophyll *a* and *b* have the same extinction coefficient). Protochlorophyll may be present in your extract but does not absorb a significant amount at this wavelength. Record your data in the chart on page 245.

CALCULATIONS

Total chlorophyll content can be determined using a millimolar extinction coefficient of 36 L/mmol-cm. Calculate the nmol of chlorophyll per gram fresh weight (see Expt. 3).

TEXT REFERENCES

Galston, A. W., P. J. Davies, and R. L. Satter, *The Life of the Green Plant* (3rd ed.), pp. 35-38. Englewood Cliffs, NJ: Prentice Hall, 1980.

Goodwin, T. W. and E. I. Mercer, *Introduction to Plant Biochemistry* (2nd ed.), pp. 465-477. Oxford: Pergamon Press, 1983.

Salisbury, F. B. and C. W. Ross, *Plant Physiology* (4th ed.), p. 391 and pp. 457-458. Belmont, CA: Wadsworth Publishing Co., 1991.

FURTHER READING

Beale, S. I., "Biosynthesis of the tetrapyrrole pigment precursor δ-aminolevulinic acid from glutamate," *Plant Physiol.* (1990), 93:1273-1279.

Castelfranco, P. A., P. M. Rich and S. I. Beale, "The abolition of the lag phase in greening cucumber cotyledons by exogenous δ-aminolevulinic acid," *Plant Physiol.* (1974), 53:615-618.

Kasemir, H., "Light control of chlorophyll accumulation in higher plants," pp. 662-686 in W. Shropshire, Jr. and H. Mohr, eds., *Encyclopedia of Plant Physiology, New Series*, Vol. 16B, *Photomorphogenesis*. Berlin: Springer Verlag, 1983.

Lew, R. and H. Tsuji, "Effect of benzyladenine treatment duration on delta-aminolevulinic acid accumulation in the dark, chlorophyll lag phase abolition, and long-term chlorophyll production in excised cotyledons of dark-grown cucumber seedlings," *Plant Physiol.* (1982), 69:663-667.

Moses, P. B. and N-H. Chua, "Light switches for plant genes," *Scientific American* (1988), 258(4):88-93.

Nadler, K. and S. Granick, "Controls on chlorophyll synthesis in barley," *Plant Physiol.* (1970), 46:240-246.

EXPERIMENT 29

Name _____

Date _____

RESULTS

Record your data in the chart below and determine the total chlorophyll content in nmol/g fresh weight.

Treatment	Weight (g)	A_{652}	Total Chlorophyll (nmol/g f.w.)

QUESTIONS

1. Explain the logic behind your choice of treatments.

2. Do your results support or refute the proposed model for chlorophyll biosynthesis? How?

Plant reactions occur at the molecular level, hence molarity is used for concentration units rather than percent or parts per million. A given level of iron, for instance, is directly comparable to nitrogen content if the concentrations are expressed in molarity, but not if they are stated in parts per million. THINK MOLAR! Familiarity with Système Internationale (SI) units, metric units and the abbreviations for fractions of molar units is essential for many of the calculations in this text. Some of the calculations require the use of constants; relevant information is included here. Both SI and traditional units are listed since both are in current use.

A. WEIGHT

SI Unit	Symbol	Grams
kilogram	kg	10^3
gram	g	1
milligram	mg	10^{-3}
microgram	μg	10^{-6}
nanogram	ng	10^{-9}
picogram	pg	10^{-12}

B. LENGTH

SI Unit	Symbol	Meters
kilometer	km	10^3
meter	m	1
centimeter	cm	10^{-2}
millimeter	mm	10^{-3}
micrometer	μm	10^{-6}
nanometer	nm	10^{-9}

C. MOLAR CONCENTRATION

SI Unit	Name	Symbol	Molar	Amount/L	Amount/mL
$10^3 \, mol/m^3$	molar	M	1	1 mol/L	1 mmol/mL
$1 \, mol/m^3$	millimolar	mM	10^{-3}	1 mmol/L	1 μmol/mL
$10^{-3} \, mol/m^3$	micromolar	μM	10^{-6}	1 μmol/L	1 nmol/mL
$10^{-6} \, mol/m^3$	nanomolar	nM	10^{-9}	1 nmol/L	1 pmol/mL

D. VOLUME

SI Unit	Common Name	Symbol	Liters
m^3	kiloliter	kL	10
dm^3	liter	L	1
cm^3	milliliter	mL	10^{-3}
mm^3	microliter	μL	10^{-6}

E. SOME ENGLISH EQUIVALENTS

SI/Metric	English
1 g	0.035 ounce (oz)
1g	0.0022 pound (lb)
1 L	1.0567 quarts (qt)
1 m	3.28 feet (ft)

F. OTHER SI UNITS, PHYSICAL CONSTANTS AND CONVERSIONS

SI unit	Symbol	Concept	Equivalents/Conversion Factors
Avagadro's number	N_A		6.02×10^{23} atoms mol^{-1}
becquerel	Bq	activity (radioactive)	$1 Bq = 1$ dis s^{-1}, $1 Ci = 3.7 \times 10^{10}$ Bq
coulomb	C	electric charge	$C = J V^{-1}$
dalton	Da		$1 Da = 1$ g mol^{-1}
faraday	F		9.648×10^4 C mol^{-1}
gas constant	R		8.31 J mol^{-1} K^{-1}
grey	Gy		$Gy = J$ kg^{-1}, $1 Gy = 100$ rd
joule	J	energy	$1 J = 4.1868$ cal
kelvin	K	temperature	$0°C = 273$ K
megapascal	MPa	pressure	$1 MPa = 10$ bar, 9.87 atm or 10^6 J m^{-3}
sievert	Sv	equivalent dose	$Sv = J$ kg^{-1}
volt	V	electrical potential	$V = J C^{-1}$
watt	W	power	$W = J s^{-1}$
		molar volume of gas	22.4 L at NTP

G. LIGHT

Type	Units	Concept	Equivalents
quantum	μmol s^{-1} m^{-2}	photon fluence rate	1μmol s^{-1} m^{-2} = 1 μEinstein s^{-1} m^{-2} = 6.02×10^{17} photons s^{-1}m^{-2}
radiometric	W m^{-2}	energy fluence rate	$1 W$ m^{-2} = 1 J s^{-1} m^{-2}

REFERENCES

Salisbury, F. T, "Système Internationale: The use of SI units in Plant Physiology," *J. Plant Physiol* (1991), 139:1-7.

Salisbury, F. T. and C. W. Ross, *Plant Physiology* (4th ed.), pp. 601-606. Belmont, CA: Wadsworth Publishing Co., 1991.

PART I. Glass pipets. It is recommended that a pipeting aid be used in conjunction with glass pipets. There are three major kinds of glass pipets, differing in speed and accuracy of use:

A. **Volumetric pipets.** The most accurate and the slowest to use, these are calibrated for one volume only and to drain dry. There is usually a bulb in the middle and a single calibration mark. The ideal procedure with volumetric pipets is as follows:

　　1. Fill to slightly **above** the calibration mark. Remove from the reagent bottle and wipe off any excess fluid from around the tip.
　　2. Place the tip against the side of the original reagent bottle. Allow the fluid to drain down to the calibration mark. Remove the pipet from the reagent bottle and **do not wipe the tip.**
　　3. Put the tip of the pipet into a new container and allow the pipet to drain, keeping the side of the tip against the wall of the container. **Do not blow out the last drop.**

B. **Ordinary pipets.** These will be calibrated along the side, but not down to the very tip. Accuracy is more difficult to achieve than with a volumetric pipet, and requires lining up the meniscus with a definite starting and ending point. Procedure:

　　1. Fill the pipet to slightly beyond the zero mark. Remove it from the reagent bottle, and wipe off the fluid from around the tip.
　　2. Return the pipet to the original reagent bottle. Put the tip against the side of the bottle, and allow it to drain until the meniscus is exactly at the zero mark. Remove from the reagent bottle, and **do not wipe the tip.**
　　3. Put the tip of the pipet into the new container and hold it against the side. Allow the fluid to drain until the meniscus is exactly at the mark for the desired quantity.

C. **Serological pipets** ("blow-out" pipets). These are the most rapid and convenient to use, at the cost of some accuracy. They can be distinguished by a ground-in ring around the mouthpiece end, and by the fact that the markings continue down to the very tip.

　　1. Same as for ordinary pipets.
　　2. Same as for ordinary pipets.
　　3. Put the tip of the pipet into the new container. **Without** touching the tip to the side, allow the fluid to drain out of the pipet completely. When it is all drained out, wait about 2 or 3 seconds, then blow out the very last drop at the bottom of the pipet. Remove the pipet without touching the tip to the sides of the container.

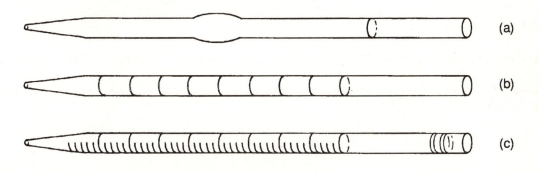

FIGURE B-1. Three types of glass pipets: (a) volumetric, (b) ordinary and (c) serological (blow-out).

Notes on Accuracy: Except with the volumetric pipets, a standard deviation of 1% is surprisingly hard to achieve. If available, pick the type of pipet suited to the accuracy desired. The greater accuracy will be obtained when the total volume of the regular or serological pipet is only a little larger than the volume of fluid to be dispensed. Remember that the pipets are calibrated for a definite temperature, usually 20°C; at other temperatures the bore of the glass changes in cross-sectional area.

Safety: The use of a rubber bulb pipet aid, Pi-pump pipetting aid or an automatic pipet is recommended at all times. **Never** pipet poisonous or radioactive chemicals by mouth.

PART II. Automatic pipettors: Relatively inexpensive automatic pipettors are now available which can accurately dispense small quantities of fluid. These may be specific for a single volume or may be adjustable. Adjustable volume pipettors must be set to the proper volume before a solution is dispensed. Appropriately sized disposable pipet tips are required for proper use of automatic pipettors. Some pipettors may be of the "double stop" variety, which require extra care for proper use, but allow retained droplets to be "blown out" of the tips.

A.   **Regular single stop automatic pipettors:**

1.   Depress the plunger and place the pipet tip into the solution to be pipetted. Release the plunger slowly, especially when pipeting large volumes. Check to see that the solution has been taken up into the pipet tip, but has not receded from the tip end. If there are bubbles present or the solution does not fill the tip to the tip end, depress the plunger and start again.
2.   Wipe the tip if a droplet is on the outside of the tip. Place it against the inside of the new container. Depress the plunger and allow the fluid to drain down the side of the container.

B.   **Double stop automatic pipettors:**

1.   Depress the plunger slowly to determine the position of the two stop points. When you feel confident of these two positions, depress the plunger to the **first** stop point and place the only the very end of the pipet tip into the fluid to be pipetted. Release the plunger slowly and follow the precautionary advice for single stop pipettors above.
2.   Wipe the tip only if necessary. Place it into the solution in the new container. Depress the plunger to expel the fluid, then take up some of the solution into the pipet tip. Then, depress the plunger all the way to the **second** stop to expel the fluid from the pipettor. Remove the tip from the solution before releasing the plunger.
3.   Dispose of the tip.

Spectrophotometric devices enable the user to determine the absorbance (or optical density) of a solution, by measuring the amount of light that passes through the solution. The three types of spectrophotometric instruments described below are the most commonly available in plant physiology laboratories.

PART I. Conventional Spectrophotometers and the Spectronic 20. The principle of the single-beam type of spectrophotometer is illustrated in Fig. C-1. Polychromatic light is passed through a prism which focuses a single wavelength of light on a slit, allowing only one wavelength of light to be transmitted. This monochromatic light then passes through the sample. The absorbance of the sample is determined by the amount of light reaching a detector, or phototube, on the other side compared to the amount reaching the detector when no sample is present ($A = \log I_o/I$). The prism may be adjusted to allow any single wavelength of visible light to pass through the sample. This kind of instrument is excellent for measuring absorbance at a single wavelength, but is not convenient for determining entire absorbance spectra, since the prism must be readjusted and the zero reset for each wavelength. The following instructions are general in nature. Be sure to follow any special advice from your instructor for the instrument you are using.

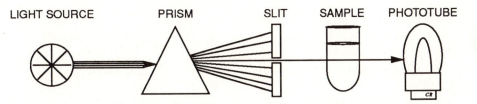

LIGHT SOURCE   PRISM   SLIT   SAMPLE   PHOTOTUBE

FIGURE C-1. Simplified diagram of the conventional single beam spectrophotometer.

Operating Procedures for the Spectronic 20.

A. **Apply power.** Turn the Power Switch clockwise to turn on the instrument. Allow the instrument to warm up for five minutes.

B. **Select the wavelength.** Turn the Wavelength Control to the proper wavelength setting as indicated on the wavelength scale.

C. **Set the zero.** Adjust the Zero Control (Power Switch) so the meter reads zero percent transmittance (the switch next to the meter should be in the up position on digital readout models). There should be no test tube in the instrument and the sample holder cover must be closed.

D. **Standardize the light control.**
    1. Fill a Spectronic 20 tube about half full of distilled water or other reference liquid. Wipe the outside of the tube to make sure that there are no fingerprints or marks which might obstruct the light path.
    2. Insert the tube into the sample holder and close the cover. Be sure to align the etched mark on the test tube with the line on the sample holder.
    3. Adjust the Light Control as required until the meter reads 100% transmittance.

E. **Make a sample measurement.**
    1. Fill another tube about half full of the sample liquid. Wipe the outside of the tube to make sure that there are no fingerprints or marks which might obstruct the light path.

2. Insert the sample into the sample holder. Align the tube as above. Close the cover. The Percent Transmittance (T) or Absorbance (A) of the sample may now be read directly from the meter. Flip the switch to the down position to read the Absorbance on digital readout models.

IMPORTANT: Step 4 must be repeated each time the wavelength is changed. When operating at a fixed wavelength for long periods, check occasionally for meter drift from 100% T.

PART II. The Colorimeter. The colorimeter operates on the same principle as the spectrophotometer, but is simpler in design. The polychromatic light is filtered through a colored glass filter which allows a broad band of a single color of light to pass (Fig. C-2). The colorimeter is most useful when the sample to be measured contains only one absorbance peak in the range of the filter to be used. For example, samples which contain colored dye or stain can be readily measured using a colorimeter. The data is obtained as colorimeter units; for the Klett-Summerson colorimeter, absorbance (optical density) may be approximated by multiplying the number of Klett Units by 0.002.

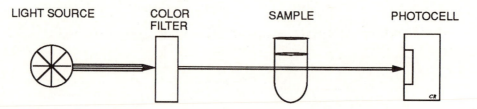

LIGHT SOURCE   COLOR   SAMPLE   PHOTOCELL
                      FILTER

FIGURE C-2. Simplified diagram of the colorimeter.

Operating Instructions for the Klett-Summerson Colorimeter.

A. **Apply power**. First, adjust the dark current using the knob on the top of the front housing to bring the red needle to the centerline. Place a Klett tube containing water (or other reference liquid) in the appropriate slot. Make sure the photocell switch on the side of the colorimeter is turned **off**, then turn on the lamp using the switch on the lamp housing at the rear. Allow about 5 minutes warm-up time.

B. **Set the zero.**
   1. Turn the scale to zero using the large knob at the front of the colorimeter.
   2. Switch on the photocell (put the switch at the side of the colorimeter in the up position). If the needle is deflected, adjust it to the center line using the zero adjustment knob at the rear left.
   3. Switch off the photocell and remove the blank test tube.

C. **Make a sample measurement**.
   1. Insert a test tube containing your sample.
   2. Turn the photocell on and adjust the red needle to the center line using the large knob at the front. Read and record the dial setting in Klett Units.
   3. Turn the photocell **off** and remove your sample. Repeat this step for each sample. There is no need to reset the zero between every measurement, but if a large number of samples are being read, then the zero should be checked occasionally.

PART III. The Diode-Array Spectrophotometer. Diode-array spectrophotometers allow the user to measure absorbance at many wavelengths simultaneously. In this case, polychromatic light passes through the sample first (Fig. C-3), then is dispersed by a prism (or "polychromator") which separates the light into narrow bands of specific wavelength. The light next reaches a bank of diodes (the "array"). Each diode is dedicated to measuring light at a narrow band of the spectrum. The absorbance at each wavelength is determined by comparing the amount of light reaching each diode to the amount reaching each diode when no sample is present (a blank). As these spectrophotometers are commonly interfaced with computers, no operating instructions are given here. Your instructor will provide the information needed if a diode-array spectrophotometer is available in your laboratory.

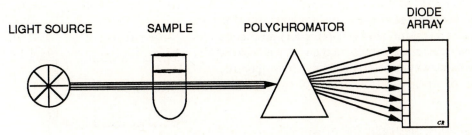

FIGURE C-3. Simplified diagram of the diode array spectrophotometer.

PART I. General Principle.

The oxygen electrode measures O_2 concentration by polarography. The electrode itself consists of 2 wires in contact with a small volume of electrolyte solution (saturated KCl), with a DC voltage (or "polarizing voltage") imposed across them by a battery. The platinum (Pt) cathode will be 0.80 volts more negative than a silver (Ag) anode. O_2 atoms, arriving at the surface of the negative Pt electrode, are reduced by electrons driven by the polarizing voltage. A corresponding number of Ag atoms at the surface of the electrode give up their electrons to the battery and become Ag^+ ions. The Ag^+ ions and Cl^- ions from the solution immediately precipitate onto the electrode as AgCl, which is insoluble. The net result is a complete flow of electrons around the circuit from the battery to the solution and back again to the opposite side of the battery (Figure D-1).

A microammeter (or a millivoltmeter placed in combination with a resistor, such as a recorder) positioned along the circuit can measure the amount of this current flow. The amount of current produced depends upon the concentration of oxygen at the surface of the platinum wire.

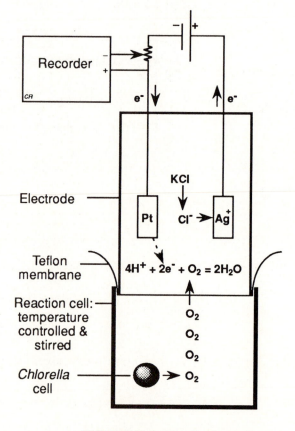

In the Clark electrode, the Pt and Ag wires and the connecting KCl solution are located behind a Teflon membrane. Neither water nor the compounds soluble in water (including many gases) will penetrate a plastic membrane of this sort. Oxygen **can** penetrate, however, and once it has entered the electrode, its reduction is sufficient to produce measurable currents. Although many compounds might be reduced by a negative voltage, specificity is obtained by using a particular bias voltage; 0.80 volts is not enough for some materials, but sufficient for oxygen.

It should be noted that the actual concentration of oxygen at the surface of the platinum is dependent upon the diffusion equilibria along the physical pathway for oxygen molecules (from the solution in the reaction chamber into and through the membrane) and the consumption of oxygen molecules at the surface of the platinum. Optimum response is not obtained unless the solution in the reaction chamber is stirred (this keeps the diffusion pathway in the external solution at a minimum). Also, both the solubility of oxygen in water and the rate of diffusion of O_2 will change if the temperature is altered during the course of the experiment.

FIGURE D-1. The O_2 electrode.

PART II. Special Precautions.

A.   The solution must be stirred continuously to keep a fresh supply of O_2 at the tip of the electrode. If stirring stops or is omitted, the current will drop rapidly.

B.   The temperature must be held constant. Distilled water used during calibration must be kept at the correct temperature. In addition, the temperature of the sample should be allowed to equilibrate before measurements are begun.

C.   The reaction cell must be filled with solution, with **no bubbles**. A bubble would serve as a reservoir of air, and the oxygen produced in your reaction could diffuse into the bubble, lowering the apparent amount produced and measurable in solution.

D.   The teflon membrane must be intact. A torn membrane will leak and cause chattering of the marker on the recorder.

PART III. Solubility of O_2 in Water.

Temperature (°C)	α^*	M Concentration $(\alpha/22.4)$	mM Concentration in equilibrium with air** $(\alpha/22.4 \times 0.20 \times 1000)$
0	0.04889	0.00218	0.4365
5	0.04287	0.00191	0.3828
10	0.03802	0.00170	0.3395
15	0.03415	0.00152	0.3049
20	0.03100	0.00138	0.2768
25	0.02831	0.00126	0.2528
30	0.02608	0.00116	0.2329
35	0.02440	0.00109	0.2178

* The solubility coefficient (α) is defined as the ratio of the volume of gas to one volume of water when the gas pressure is 760 mm Hg (i.e. at an O_2 pressure of 0.1 MPa).

**To determine the concentration of O_2 in a solution which is in equilibrium with air, multiply by 0.20 since air is 20% oxygen.

Adapted from Unbreit, W.W., R.H. Burris and J.F. Stauffer, *Manometric and Biochemical Techniques* (5th ed.), p.62, Burgess Publ. Co., Minneapolis, 1972.

PART IV. Schematic Design of Circuit Box for Setting Polarizing Voltage.

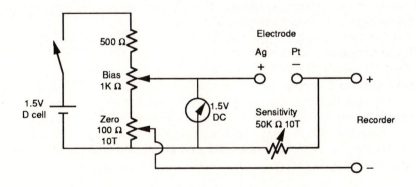

PART I. The Nature of Radioactivity.

Particles are emitted from disintegrating (radioactive) atoms. Three kinds of natural radioactive particles occur:

alpha (α) particles = positively charged helium nuclei, moving at high speed.
beta (β) particles = electrons, or positrons, moving at high speed.
gamma rays = X-rays = (uncharged) electromagnetic radiation, similar to light but of a shorter wavelength, and higher energy per photon.

The isotopes usually used in laboratory work give off either beta or gamma radiation. Each kind of radiation consists of particles (electrons or photons) which have a definite energy level per particle.

The most important characteristic of radioactive emissions is that as they speed past atoms or molecules, they interact by knocking electrons out of the atoms. The atoms or molecules in turn acquire an electric charge, and become **ions**. Radioactive emissions are therefore known as **ionizing radiations**. With each ionization act the radioactive particle loses some of its energy, which is gradually consumed. The more energetic the gamma ray or beta particle in the first place, the farther it can penetrate through matter before losing its ability to cause ionizations. Thus, high energy radiations penetrate more deeply than low energy ones.

Biological damage results from direct or indirect damage to sensitive molecules as they become ionized. Thus the biological effectiveness of radiation depends on the intensity of ionizations it produces.

PART II. Units of Radioactivity.

The SI unit for radioactivity is the becquerel (Bq), which is equivalent to 1 disintegration per second. However, the more commonly used designation is the curie (Ci):

1 Ci   = 3.7×10^{10} disintegrations per second (dps) or

= 2.22×10^{12} disintegrations per minute (dpm)

One millicurie (mCi) is the above x 10^{-3}; one microcurie (μCi) is the above x 10^{-6}. One Ci is equal to 3.7×10^{10} Bq.

The recorded counts per minute (CPM) will depend on the proportion of the total number of disintegrations that your instrument is able to record. Thus, if your instrument measures 10 CPM when your sample has 100 DPM, your instrument has an efficiency of 10%.

PART III. Half-life Considerations.

Each radioactive isotope differs in the rate at which it decomposes. This is a statistical matter. Any one atom disintegrates only once, but in a large population of such atoms, the moment of disintegration of each one will occur in a random manner. The probability of disintegration is constant with time and, for any given mass of an element, the time for 50% decomposition is constant. This constant is called the half-life of the element. Half-lives of some of the commonly used elements are: 5,730 years for ^{14}C, 12.26 years for ^{3}H, 87 days for ^{35}S and 14.3 days for ^{32}P.

PART IV. Absorbed Radioactivity.

The unit of absorbed radioactivity is the rad (rd) or Gray (Gy). The absorbed dose quantifies radiation exposure on a physical basis. It is the amount of energy a mass of material absorbs when exposed to radiation. The SI unit is the Gray which is equal to 100 rd or 1 J/kg.

PART V. Dosage of Radioactivity.

The dosage of radioactive emissions is defined by the effects the radiation has, rather than by the number of particles passing by; i.e., it is defined in terms of the number of ion pairs produced in a given volume or mass of matter, per unit time. The number of alpha, beta or gamma particles it takes to do this will depend on the energy of the particle. The effect of a given energy particle also depends on the nature of the material it is passing through. The unit for dose equivalent, which quantifies radiation exposure on a biological basis, is the rem or Sievert (Sv). The SI unit is the Sievert, which is equal to 100 rem or 1 J/kg.

With a given radiation dosage at the surface of the skin, the actual tissue dosage will decrease for successively deeper levels (because the radiation becomes less energetic as it penetrates and creates ions in the upper levels). The tissue dosage tends to equal or be lower than air dosage, as you pass more than 5 cm below the surface.

The dose a person receives depends on:

1. The activity of the source.
2. The nature of the source: An isotope may emit one or several characteristic radiations with each decay.
3. The shape of the source: The shape and strength of the radiation field will vary according to the shape of the source. The shape of the source and the position and nature of nearby material also determine the amount of scattered radiation, which may significantly add to the dose.
4. The exposure time at a given dose rate: dose rate x exposure time = total dose.
5. The distance of the target from the source. **Inverse square law**: The strength of a radiation field falls off as the square of the distance from the target to the source.
6. The material between the target and the source, including the solution in which an isotope may be found. The ability of the material to stop or attenuate radiation depends on the probability of interaction, the density of the material and the nature of the radiation. Beta particles are most safely stopped by low density materials. Two important examples:
    a.   For ^{14}C (a beta particle emitter) the maximum range in air is about 24 cm, while the maximum range in water is only 0.3 mm.
    b.   For ^{32}P (a beta particle emitter) the maximum range in air is about 600 cm, while the maximum range in water is only 0.8 cm.
    Because of these differences, Plexiglas shielding is often used for larger amounts ($> 10\,\mu Ci$) of ^{32}P but not for ^{14}C.
7. The radiological half-life of the isotope and the biological half-life of the compound, if the exposure is over a period of time and/or is internal.

PART VI. Some Practical Examples.

Remember that radiation intensity falls off inversely by proportion to the square of the distance to the source.
1. A dental X-ray is approximately equivalent in effective exposure (tissue damage) to holding $10\,\mu Ci$ of ^{32}P (120 cps on the counter at one foot range) one foot from your jaw for several hours. Thus a "hot spot" of $0.1\,\mu Ci$ of ^{32}P on a surface, registering 90 cps on the counter at point blank range, even if brought within one foot of the body causes negligible amounts of radiation damage (less than 1% of a dental X-ray).

2. As an important contrast to the first example, consider the same $0.1\,\mu Ci$ of ^{32}P distributed on one square centimeter on your skin. This small amount of radioactive solution will give the small volume of tissue

immediately underlying the skin (which ^{32}P readily penetrates) a dose rate of about 1000 mrem/h. Thus, in a little more than one hour that tissue will have received the same dose that it would have received if the whole hand had been exposed to the maximum permissible 1500 mrem/week.

PART VII. Why We Use Radioactive Isotopes.

In plant physiology, as in other kinds of experimental biology, radioactive isotopes are extremely useful in elucidating reaction pathways, studying uptake mechanisms, etc. This is because the radioactive isotope shares the chemical and biological characteristics of the common, non-radioactive element, but produces emissions which permit the measurement of far less material (10 to 1000 times less, depending on the efficiency of assay for the non-radioactive material.) The proper handling and understanding of ways to assay radioactive materials are therefore important tools for the biological sciences.

The radioactive isotopes used most commonly are those of C, H, P, S and I. Unfortunately, none are available for N or O, and the half-life for K is so short as to make it almost unusable.

PART VIII. Safety Precautions.

IMPORTANT: Great care must be taken when using radioactive isotopes to avoid transferring any isotopes to your hands, clothes, or equipment. The following precautions are mandatory when a high-energy β radiation emitters like ^{32}P and ^{14}C are used. PLEASE OBSERVE THE FOLLOWING PRECAUTIONS FOR THE HANDLING OF RADIOACTIVE MATERIALS DURING CLASS!

1.   Never pipet radioactive solutions by mouth. Automatic ipetting devices must be used to transfer radioactive solutions.
2.   Wear latex gloves and a disposable lab coat when radioactive solutions are used.
3.   Label all materials which come in contact with the radioactive solution with warning tape. Remove the tape when these items are decontaminated.
4.   Carry out all work in a tray lined with and placed on absorbent paper so that minor spills will be contained and radioactivity will not be spread around the laboratory. Also keep the stock solution in secondary containment.
5.   Never hold radioactive material near one's eyes. The eye is especially sensitive to ionizing radiation. As an isotope with higher energy than ^{14}C ^{32}P is used than when is used.
6.   Dispose of waste radioactive materials in the special containers provided, but do not put other, non-radioactive materials, in these containers.
7.   Tell your instructor of any accidental spillage of radioactive solution!
8.   Monitor all materials, equipment, and your bench during and after use. The instructor will monitor your hands and shoes before you leave class today to ensure that no radioactive materials are carried out of the classroom.
9.   Always keep solutions containing ^{32}P behind Plexiglas shielding when quantities greater than 10 μCi are used. Shielding is suggested for all use of ^{32}P; practice with non-radioactive materials is strongly recommended before using shielding with radioactive materials.

PART IX. How We Measure Radioactivity.

1.   **Autoradiography**. A sample containing the radioactive isotope is placed in intimate contact with X-ray film. The ions produced by radiation accomplish chemical work leading to decomposition of the silver grains in a photographic or X-ray emulsion. Autoradiography is an excellent method for detecting ^{14}C.

2.   **Geiger-Müller counting**. The radioactive ray enters a chamber containing an ionizable gas. The ions resulting from collision of the particle with the gas are then attracted to, and affect the electrical properties of, a positively charged electrode. The resulting electrical pulses are counted by the electronic system of

the machine. A disadvantage is that only those emissions that are traveling in the direction of the chamber and are energetic enough to enter the chamber can be counted. This method is good for measuring ^{32}P, but is not efficient for ^{14}C, because the β radiation emitted by ^{14}C is not sufficiently energetic to pass through the "window" at the end of the chamber.

3. **Liquid scintillation counting.** The radioactive sample is placed in intimate contact with special molecules called "fluors" in a solution. When a fluor absorbs radioactive energy it emits a quantum of visible light. The light is absorbed by a bank of surrounding, highly sensitive photomultipliers. Scintillation counting is excellent for ^{3}H, ^{14}C and ^{32}P.

PART X. Collection and Disposal of Radioactive Materials.

^{14}C: Collect liquid radioactive waste in plastic containers which are placed in secondary containers. Solid waste may be collected in properly labeled metal cans lined with plastic bags. Scintillation vials containing radioactive materials may be collected in their original boxes.

Radioactive waste containing ^{14}C must be disposed of according to regulations for low-level radioactive emitters in your state. Follow the instructions of your local environmental safety office.

^{32}P: Collect liquid radioactive waste in plastic or glass containers which are placed in secondary containers. Plastic carboys may be reused after an appropriate period of decay and the liquid has been properly discarded. Solid waste may be collected in properly labeled metal cans lined with plastic bags. No sharp objects or "radioactivity" labels should be placed in ^{32}P waste as it will eventually be deposited with regular trash.

Since the half life of radioactive phosphate is only 14.3 days, the radioactive emissions will be reduced to background levels after about five months. ^{32}P radioactive waste may be stored in properly labeled containers in a restricted area until this time. The waste may then be placed in regular trash containers. Follow the regulations of your state and institution.

FURTHER READING

Appling, J., *Working With Radiation: Answers to Your Questions about Practical Issues*. Ithaca, NY: Office of Environmental Health, Cornell University, 1986.
Shapiro, J. *Radiation Protection: A Guide for Scientists and Physicians* (3rd ed.). Cambridge: Harvard University Press, 1990.

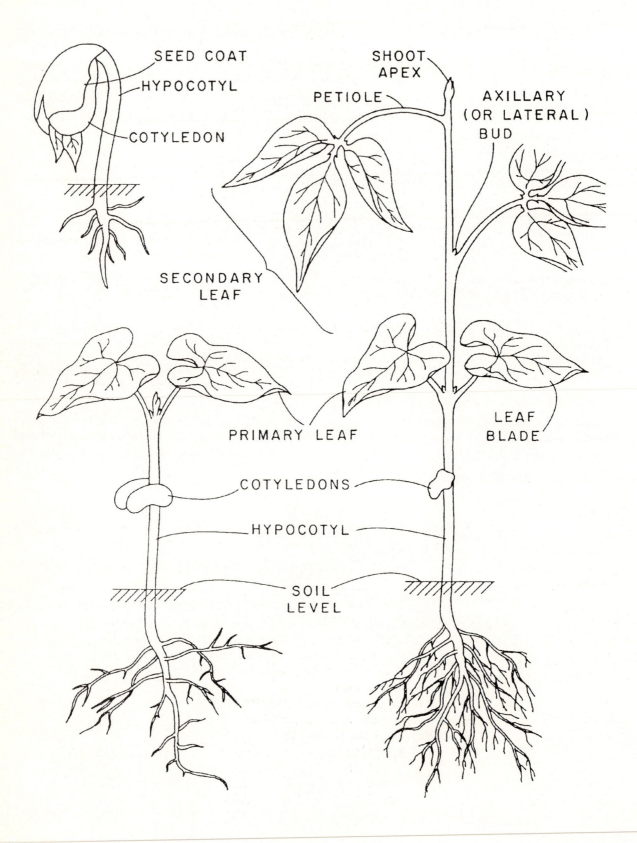

SEED COAT

HYPOCOTYL

COTYLEDON

SHOOT APEX

PETIOLE

AXILLARY (OR LATERAL) BUD

SECONDARY LEAF

PRIMARY LEAF

LEAF BLADE

COTYLEDONS

HYPOCOTYL

SOIL LEVEL

APPENDIX G

**Equipment, Solutions and
Plant Material**

Sources for special equipment, chemicals and plant material are listed for each experiment. Sources of equipment are listed only if the item is difficult to locate. Instructions for making solutions are included, along with the recommended amounts for a class of twelve students working in groups of two. In most cases, simply multiply by the number of laboratory sections to determine the total amount required. For some solutions (e.g. inhibitors like DCMU), where it is impractical to make the small amounts needed for each class, the amount indicated will be sufficient for a number of classes. Some solutions **must** be freshly made and are so marked.

EXPERIMENT 1: Determination of the Ascorbic Acid Content of Cabbage

Solutions:

4000 mL	5% metaphosphoric acid: 200 g/4000 mL.
1000 mL	Dichlorophenol-indophenol (DCIP): 0.8 g/1000 mL. Dispose of properly.
50 mL	Ascorbic acid (4.0 mg/1.0 mL): 0.20 g/50 mL. Cover with aluminum foil and refrigerate. Prepare just before class as ascorbic acid is readily oxidized in aqueous solution.

Plant material: Green cabbage may be purchased at the supermarket; one head is sufficient per class. Virtually any green vegetable or citrus fruit may be used instead.

Equipment: Miracloth is available from Calbiochem, P.O. Box 12087, San Diego, CA 92112-4180.

Alternatives: Use other fruits or vegetables (but not red cabbage!) and compare to cabbage. Compare ascorbic acid content of leaves to that of the stem. Boil cabbage for different lengths of time.

Class time: 3 to 3.5 h.

EXPERIMENT 2: Amylase: Enzyme Assay

Solutions:

500 mL	10 mM citric acid-sodium citrate buffer at pH 5. Make 500 mL 10 mM sodium citrate (1.47 g/500 mL) and 250 mL 10 mM citric acid (0.53 g/250 mL). Keep refrigerated.

To make buffer at pH 5.0 combine:

$$325 \text{ mL} \quad 10 \text{ mM sodium citrate}$$
$$\underline{+175 \text{ mL} \quad 10 \text{ mM citric acid}}$$
$$500 \text{ mL} \quad 10 \text{ mM citric acid-sodium citrate buffer at pH 5.0. Check pH.}$$

1500 mL	1 N HCl: 124.5 mL concentrated HCl made up to 1500 mL (add acid to water!).
500 mL	Iodine solution: 2.5 g KI, 0.18 g KIO_3, 500 mL 2 mM NaOH (0.04 g/500 mL).
500 mL	0.05% starch solution in 0.05 M citric acid-sodium citrate buffer at pH 5.0 and
100 mL each	0.05% starch in the above buffer at pH 3, 4, and 6.

Make the following stocks:
500 mL 0.05 M sodium citrate: 7.35 g/500 mL
500 mL 0.05 M citric acid: 5.25 g/500 mL
To make the buffer (check and adjust pH):

pH	Sodium Citrate	Citric Acid	Total Amount
3	6.3 mL	83.7 mL	90 mL
4	31.0	59.0	90
5	265.5	184.5	450
6	72.9	17.1	90

To make stock 0.5% starch solution: BOIL 0.5 g soluble starch/100 mL water. Check the volume after boiling. Add 10 mL of this starch solution to 90 mL of each buffer at pH 3, 4 and 6. Add 50 mL to the pH 5 buffer. Save the remaining starch solution for the solutions below. Do not refrigerate.

100 mL each 0.05% starch solution in 0.05 M phosphate buffer at pH 7 and 8. To make buffer first make stocks:

   50 mL 0.05 M KH_2PO_4: 0.34 g/50 mL.
   150 mL 0.05 M K_2HPO_4: 1.3 g/150 mL.

Then combine:

pH	KH_2PO_4	K_2HPO_4	Total Amount
7	35.1 mL	54.9 mL	90 mL
8	4.8	85.2	90

Check the pH. Add 10 mL of boiled 0.5% starch solution to each of the above to get 100 mL 0.05% starch solution at the proper pH. Do not refrigerate.

Plant material: Wheat seeds (*Triticum aestivum* var. Frankenmuth or Houser are suggested; most are acceptable), germinated between wet paper towels and covered with foil for 72 h. Other varieties of wheat, barley or oat seeds will work as well.

Class time: About 3 h.

EXPERIMENT 3: Separation, Identification and Quantitation of Plant Pigments

Solutions:

50 mL Petroleum ether-acetone-chloroform (3:1:1). Use AR grade chemicals. Store in a fume hood. Should be freshly made. Dispose of properly as a chlorinated solvent. In general, avoid mixing solvents with chlorinated solvents for disposal.

100 mL 100% acetone in a Repipet set at 4 mL.

100 mL 90% acetone for pigment extraction: 90 mL of acetone plus 10 mL of distilled water. NOTE: If 10 mL of 0.1 N NH_4OH is added instead of water, the amount of pheophytin formed from the acid induced breakdown of chlorophyll will be greatly decreased. Oxalic acid present in the vacuoles of spinach induces the breakdown; the added NH_4OH maintains the extract at a higher pH. Both methods offer pedagogical opportunities.

10 mL Spinach chlorophyll preparation. One batch will be sufficient for several classes and will keep for a week below 0°C. Grind 25 g of spinach (deveined) in 100 mL of 90% acetone in a blender. Filter through a Büchner funnel. Add 10 mL of hexane and sufficient water to make a final acetone concentration of 70% (about 20 mL of water). Place in a separatory funnel. The pigment extract will accumulate at the top. Remove the acetone-water phase and wash the remaining extract twice with distilled water. Collect the hexane portion in a vial, cover it with foil and place it in the freezer.

Plant material: Spinach (*Spinacia oleracea*) may be purchased from the supermarket. One 10-oz package will be more than enough.

Preparation: Place silica gel strips in an 80°C oven at least one hour before class time.

Class time: Less than 3 h., especially if a Diode Array Spectrophotometer is available. This experiment may be coupled with another such as Starch Production or Greening.

EXPERIMENT 4: Whodunit - or - the Influence of Light Intensity on Starch Production in Photosynthesis

Solutions:

2000 mL   95% ethyl alcohol.

500 mL   Iodine stain. Dissolve 1 g of iodine in a few mL of ethyl alcohol, then add 10 g KI. Bring
                 to 500 mL with distilled water. Store in a brown bottle or cover with foil.

Plant material: Red Kidney bean (*Phaseolus vulgaris* var. Redcloud) plants, 2 weeks old. The experiment
                 works best if the plants have prominent primary leaves but undeveloped secondary leaves.
                 Place the plants in the dark the day before the experiment.

Alternatives: Paint a portion of the leaf with DCMU or use a stencil to create a design. Variegated
                 leaves may be used as an interesting alternative.

Class time: Five to ten minutes set-up time is needed before the 90 minutes of light exposure.
                 Development of the prints takes about 10 minutes. May be completed in conjunction with
                 another experiment, such as Plant Pigments.

EXPERIMENT 5: Photosynthesis: Partial Reactions in Cell-Free Preparations

Solutions: All flasks and bottles for this experiment should be labeled for Part I, II or III. Several of the
                 inhibitors may be used for more than one Part.

Part I.

500 mL   DCIP Reaction Mixture (blue):

                 0.07 mM DCIP: 0.01 g/500 mL.

                 100 mM KCl: 3.7 g/500 mL.

                 20 mM Tricine: 1.8 g/500 mL.

                 Make up as one solution, adjust the pH to 7.5, then adjust the volume. Check absorbance
                 and dilute if too dark for reading on the spectrophotometer. Dispose of properly.

100 mL   0.01 mM dichlorophenyl-dimethylurea (DCMU). Make up 10 mM stock solution (0.23
                 g/100 mL). Bring 0.1 mL of the stock to 100 mL with distilled water. Use for Parts II and
                 III also. Stock solution may be refrigerated for long-term storage. Dispose of properly.

100 mL   30 mM NH_4Cl: 0.16 g/100 mL. Use for Parts II and III also.

Part II.

100 mL   Ferricyanide solution (yellow):

                 0.03 g K ferricyanide.

                 0.74 g KCl.

                 0.36 g Tricine.

                 Make as one solution in about 90 mL, adjust the pH to 7.5 and bring to 100 mL. Dispose
                 of properly.

Part III.

250 mL   Basal solution for phosphorylation:

                 Prepare a double concentration basal solution:

                 200 mM Tricine: 8.96 g/250 mL.

                 200 mM NaCl: 2.9 g/250 mL.

                 20 mM $MgCl_2 \cdot 6H_2O$: 1 g/250 mL.

                 20 mM isoascorbic acid: 0.9 g/250 mL.

                 Make up as one solution. Measure out 125 mL of the above and add 0.75 mL of 1 M
                 KH_2PO_4 stock (1.36 g/10 mL). Adjust the pH to 8.3 before bringing to 250 mL. This
                 solution should be freshly made. (Use the remaining 125 mL of double strength Basal
                 Solution for the phosphate solutions that follow.)

Final concentrations (at pH 8.3):
    100 mM Tricine.
    100 mM NaCl.
    10 mM $MgCl_2$.
    10 mM isoascorbate.
    3 mM KH_2PO_4.

100 mL each   Phosphate solutions:

Place 25 mL of the **double** concentration basal solution into each of five 125-mL flasks and add the proper amounts of 1 M KH_2PO_4 as follows:

Label	Amount of KH_2PO_4
0 mM KH_2PO_4	0.0 mL
0.5 mM KH_2PO_4	0.05
1.0 mM KH_2PO_4	0.1
1.5 mM KH_2PO_4	0.15
2.0 mM KH_2PO_4	0.2

Bring each to 90 mL and adjust the pH to 8.3. Adjust to a final volume of 100 mL.

100 mL   15 mM ADP (pH 8.0): 0.64 g/100 mL. Be careful adjusting the pH, as overshooting the pH will ruin the ADP. Add 10 M KOH until pH 7 is reached, then use a lower concentration of KOH. Freeze in 10 mL lots.

100 mL   10% trichloroacetic acid (TCA): 10 g/100 mL. Be careful!

100 mL   0.5 mM phenazine methosulfate (PMS): 15 mg/100 mL. Wrap in foil and keep cold.

400 mL   Phosphate Developing Reagent (freshly made). Place 40 mL of 10 N H_2SO_4 (add 20 mL concentrated sulfuric acid to 20 mL distilled water) in a flask on a stirrer in a fume hood. Add 6.4 g ammonium molybdate (to make a **final** concentration of about 1% molybdate) and let dissolve. Add 20 g ferrous sulfate and 360 mL distilled water to the 40 mL acid-molybdate mix. The solution should be light brown in color. Dispose of this dilute molybdate containing solution according to your local regulations.

For chloroplasts: Make the following solutions in advance:

500 mL   Grinding and suspension buffer (SCHB). Prepare in about 450 mL:
    400 mM sucrose: 68.5 g/500 mL.
    200 mM choline Cl: 14 g/500 mL.
    20 mM HEPES buffer, pH 7.8: 2.4 g/500 mL.
    2 mg/mL bovine serum albumin (fatty acid free): 1 g/500 mL.

Cool to 0°C, then adjust to pH 7.8. Bring to volume.

100 mL   80% acetone.

Make the thylakoids daily just before class; 15 mL of thylakoids at 0.5 mg/mL for a class of twelve.

1)   Devein spinach then weigh out 30 g of leaves. Wash and blot dry. Grind the leaves in a cold blender with about 100 mL of cold SCHB for a minimal amount of time (e.g. 10 seconds).

2)   Pour through 8-10 layers of cheesecloth (squeeze) into a chilled beaker (on ice). Pour into two centrifuge tubes and spin for 5 minutes at 1900 g in a refrigerated centrifuge.

3)   Discard supernatant and resuspend each pellet in 2 mL of SCHB. Combine in one tube; rinse the empty tube with 25 mL SCHB buffer and add it to the tube containing the thylakoid preparation. Put more buffer in the empty tube for balance and spin for 5 minutes at 1900 g.

4)   Resuspend the thylakoids in 5-10 mL of SCHB. Keep on ice.

5)   Determine the chlorophyll concentration as follows: Place 0.1 mL thylakoid suspension in a 10-mL volumetric flask and bring to volume with 80% acetone. Sediment this solution in a clinical centrifuge for 2 minutes (cork the test tubes). Sediment a second time if the supernatant is cloudy.

Read the absorbance of the supernatant at 645 nm and 663 nm.

$$\# \mu g \text{ chlorophyll/mL} = 20.2 \times A_{645} + 8.02 \times A_{663}$$

Multiply by 100 to get the # μg/mL of the original suspension. Multiply by the #mL of thylakoid suspension to get total # μg chlorophyll. To determine the #mL of final suspension at 500 μg/mL divide by 500. Dilute by raising the volume to the determined number of mL.

Plant material: Spinach (*Spinacia oleracea*) may be purchased at the supermarket. One package will make about three batches.

Preparation: It is helpful to test the standard curve before class to ensure that 0.5 mL of the [Pi] in basal solution diluted to 1.0 mL will match the [Pi] in tube 11 when stained with the colorimetric reagent.

Information about O_2 electrode is found in the section on Whole Cell Photosynthesis (Expt. 8) and in Appendix D.

Glassware used in the phosphorylation experiment (Part III) may be prewashed with a phosphate free detergent (such as Triton X-100), then rinsed first with a dilute solution of nitric acid and finally with distilled water to reduce phosphate contamination.

Alternatives: Add ADP and Pi to the reaction mix in Part II. Each part of the experiment is a complete experiment in itself; any part may be omitted.

Class time: 3-3.5 h. It is recommended that student groups complete different parts (e.g., I and III or I and III).

EXPERIMENT 6: Carbon Fixation

NOTE: The use of radioactive materials requires a permit. Check with your college administration for permission to use ^{14}C.

Solutions:

5 mL	0.2 M ATP: 0.55 g/5 mL. Adjust pH **carefully** to 7.0. Keep frozen.
4 mL	0.2 M ribose-5-phosphate (disodium salt): 0.22 g/4 mL. Adjust the pH to 8. Keep frozen.
10 mL	1 M $MgCl_2$: 2.03 g/10 mL.
200 mL	Suspension Medium:

0.5 mM tetrasodium EDTA: 0.038 g/200 mL.

0.5 mM dithiothreitol (DTT): 0.015 g/200 mL.

0.04 M HEPES: 1.9 g/200 mL.

Make as one solution. Adjust to pH 8 and keep cold.

Reaction Mixtures (should be freshly made and kept cold):

	Complete	-ATP	-Ri-5-P	-$MgCl_2$
Suspension Medium	18.2 mL	19.2 mL	18.7 mL	18.5 mL
1 M $MgCl_2$	0.3	0.3	0.3	0.0
0.2 M Ri-5-P	0.5	0.5	0.0	0.5

Add 1.0 mL 0.2 M ATP to each reaction mixture (except -ATP) at the beginning of class. The ATP must be kept frozen until then.

35 mL	10% perchloric acid. Combine 5 mL 70% perchloric acid with 30 mL of water. Neutralize spills with dilute aqueous NaOH; use absorbent to collect the spilled material.
10 mL	30% NaOH: 3 g/10 mL. Add the NaOH slowly. Exothermic!
300 mL	scintillation cocktail. NOTE: Premade, biodegradable scintillation cocktail is now commercially available and is recommended. The following recipe may be used but contains toluene and must be used with care. Prepare and dispense in a fume hood.

200 mL purified toluene.

0.03 g POPOP (dissolve in toluene first).

1.5 g PPO.

100 mL Triton X-100.

It is convenient to keep the cocktail in a Repipet set for the desired amount (3 mL) and stored in a fume hood. Dispose of properly.

1.6 mL $NaH^{14}CO_3$ (3.125 μCi/mL) in 0.02 M $NaHCO_3$ (0.168 g/100 mL. Make a stock solution of 0.25 mCi/5 mL H_2O and keep it frozen. The class solution should be freshly made: 2.5 μCi (0.05 mL stock)/0.8 mL 0.02 M $NaHCO_3$. Dispose of properly (see Appendix E).

500 mL Homogenization medium for chloroplast preparation. Make in advance and keep cold:

0.05 M Tris buffer: 3.03 g/500 mL.

0.01 M KCl: 0.37 g/500 mL.

0.4 M sucrose: 68.5 g/500 mL.

Prepare as one solution and adjust to pH 7.8. Freeze for long-term storage.

To make chloroplast preparation: Make just before class, one batch at a time. Have ready: cold blender, 600-mL beaker on ice, eight centrifuge tubes on ice, funnel and cheesecloth.

1) Devein and weigh spinach: 60 g/batch. Rinse with distilled water and blot with paper towels.

2) Place in a cold blender with 100 mL cold homogenization medium and grind at high speed for as short a time as possible.

3) Filter through double cheesecloth into a chilled beaker and divide into 4 centrifuge tubes.

4) Sediment at 120 g for 8 minutes in a refrigerated centrifuge.

5) Pour supernatant into 4 cold centrifuge tubes (i.e., discard the pellet and save the supernatant).

6) Spin the supernatant at 1075 g for 8 minutes in a refrigerated centrifuge.

7) Discard the resulting supernatant and resuspend the pellets (all 4) in 8 mL (total) suspension medium.

One batch makes 8 mL, which is sufficient for six groups of two students each. If more than one batch is required, each must be ground in the blender separately, but more than one batch may be centrifuged at the same time.

Plant material: Spinach (*Spinacia oleracea*) may be purchased from the supermarket. One package will be enough for about 2 batches.

Class time: 3-3.5 h.

Alternatives: Omit N_2 to determine if O_2 is competitive for Rubisco. Run the reaction at two temperatures to determine Q_{10}. Use boiled chloroplast preparation in an additional control treatment.

Cleanup: Wipe tests should be performed at lab benches where radioactive materials have been used. Moist circles of filter paper (2.5 cm) must be wiped over the area (about 100 cm^2) to be checked, then each placed in a 20-mL scintillation vial containing 5 mL of scintillation cocktail. Wipe a known "cold" area as a control. Place the filter at the bottom of the vial with the wiped side up. Count in a Liquid Scintillation Counter for 1 to 5 minutes each. CPM readings should be comparable to background levels. Areas which reveal higher counts must be carefully cleaned and wiped again.

EXPERIMENT 7: Light Relations in Whole Cell Photosynthesis

Solutions:

1000 mL	*Chlorella* medium:

	5 mL	NaFeEDTA (Use Mineral Nutrition stock solution).
	26.7 mL	1 M KH_2PO_4: 13.6 g/100 mL stock.
	7.3 mL	1 M K_2HPO_4: 17.4 g/100 mL stock.
	15 mL	1 M Urea: 6 g/100 mL stock.
	5 mL	1 M $MgSO_4 \cdot 7H_2O$: 24.6 g/100 mL stock.
	1 mL	trace elements - see below.
	10 g	dextrose: add just before use and refrigerate.

Adjust pH to 6.0-6.1.

200 mL	Trace element solution:

	10 mg	Na_2MoO_4.
	320 mg	$MnCl_2$.
	2.4 mg	$CuSO_4 \cdot 5H_2O$.
	280 mg	$ZnCl_2$.

20 mL	$20 \mu M$ dichlorophenyl-dimethylurea (DCMU). Make up 20 mM stock solution (0.23 g/50 mL). Add $10 \mu L$ of stock to 10 mL water.
250 mL	0.1 M K_2CO_3: 3.45 g/250 mL.
1500 mL	0.1 M $NaHCO_3$: 12.6 g/1500 mL.
200 mL	0.1 M K_2CO_3-$NaHCO_3$ buffer (freshly made):

$$30 \text{ mL } 0.1 \text{ M } K_2CO_3$$
$$\underline{170 \text{ mL } 0.1 \text{ M } NaHCO_3}$$
200 mL of buffer at pH 8.9. Check pH.

125 mL	*Chlorella*, prepared just before class:

1) Split a 50 mL suspension (see below) into two 25 mL lots.
2) Sediment at room temperature at 1075 *g* for 5 minutes.
3) Pour off supernatant.
4) Wash the pellets once with 0.1 M K_2CO_3-$NaCO_3$ buffer. Then use a paintbrush to resuspend in a few mL buffer.
5) Combine the two suspensions and make up to 125 mL with buffer. Save the remaining buffer for resuspending the 3X concentration *Chlorella* during class.

Plant material: *Chlorella vulgaris* var. viridis may be obtained from the American Type Culture Collection, 12301 Parklawn Drive, Rockville, MD 20852. *Chlorella* is also available from Carolina Biological Supply.

The culture is received on a "slant" of agar and may be kept in the refrigerator in this form for about 6 months. Two weeks before class, under sterile conditions, transfer a small portion of the *Chlorella* from the slant to 20 mL of sterile medium (autoclaved). Gently shake the culture in the light at room temperature for 1 week. Under sterile conditions, transfer 1 mL of this mixture to several batches (depending on the number and size of your classes) of 50 mL of sterile medium and shake at room temperature in the light for a week. It is recommended that several test runs of cell culture be performed to determine the optimum timing for growth of the cells under the conditions present in your laboratory. To make new slants, add 1% agar to fresh medium and autoclave. Put 10 mL each into 25 x 150-mm test tubes, autoclave, then allow to cool at a 45° angle. Under sterile conditions, transfer a swipe of *Chlorella* from the old slant to the new. Leave the newly inoculated slants in the light at room temperature for several weeks, then store in the refrigerator.

Equipment: YSI 5331 Clark O_2 probes and accessories may be purchased from Yellow Springs Instrument Co., Inc., Yellow Springs, Ohio 45387. Water jacketed reaction cells may be purchased from Gilson Medical Electronics, Inc., P.O. Box 27, 3000 W. Beltline Hwy., Middleton, WI 53562. The polarizing voltage may be set using the circuit box diagrammed in Appendix D (see also Fig. G-1), which may be constructed at relatively low cost. Alternatively, the YSI Biological Oxygen Monitor Model 5300 and YSI 5301 may be used, but for considerably higher cost. Oxygen electrodes are also manufactured by Hansatech and Rank Brothers. Choose a 1 mV recorder set on the 10 mV scale.

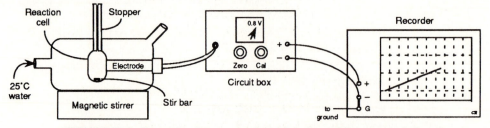

FIGURE G-1. Set-up for the oxygen electrode using a Gilson water-jacketed reaction cell and a circuit box built according to the scheme in Appendix D.

100-W incandescent flood lamps, placed about 1 foot from the reaction cell, will provide approximately 500 μE m^{-2} s^{-1} of light, which should be enough to saturate the standard *Chlorella* prep. The light may be focused through a round bottom flask containing water, which also acts as a heat filter. Light filters may be made from layers of plastic screening or with cheesecloth; the amount of light may be measured with a photon-flux meter, which may also be used to measure the light during class and replace "% light intensity."

Preparation:	A little acetic acid in the water bath will keep the water clear. It is helpful to check the electrodes before class to ensure that calibration time will be minimal, especially if the electrodes are used by more than one class. Keep some distilled water in the refrigerator and adjust it to 15°C for the second calibration. Put the *Chlorella* on ice to bring the temperature to 15°C for the low temperature light curve.
Class time:	3-3.5 h.

EXPERIMENT 8: Respiratory Control in Potato Tuber Slices and Mitochondria

Solutions:
Part I.

10 mL	10 mM CCCP in DMSO: 0.02 g/10 mL. Wear latex gloves during preparation and use; DMSO will readily penetrate the skin. Dispose of properly.
50 mL	50 mM KCN: 0.16 g/50 mL (used for Part II as well). Use care. Dispose of properly. KCN waste from Part II will be mixed with other waste liquids in an aspirator. Be sure to label for all materials; do not attempt to convert cyanide to cyanate in mixed waste.
500 mL	Resazurin solution: Make a stock solution containing 0.01 g/100 mL. Bring 25 mL of the stock to 500 mL with distilled water.
500 mL	0.5% Photo-Flo solution: Bring 2.5 mL of Photo-Flo to 500 mL with distilled water.
50 mL	Vaspar: melt together 25 mL of petroleum jelly and 25 mL of parafin. Rewarm for class.

Part II.

500 mL	Homogenization mix: make in 30 mM MOPS buffer (3.14 g/500 mL), pH 7.5. Keep cold.
	0.3 M mannitol: 27.3 g/500 mL.
	1 mM tetrasodium EDTA: 0.19 g/500 mL.
	0.1% bovine serum albumin (BSA, fatty acid free): 0.5 g/500 mL.
	4 mM cysteine: 0.24 g/500 mL.

100 mL	Washing medium:
	0.3 M mannitol: 5.4 g/100 mL.
	1 mM tetrasodium EDTA: 0.4 g/100 mL.
	0.1% BSA (fatty acid free): 0.1 g/100 mL.
	in 10 mM MOPS buffer (0.21 g/100 mL), pH 7.4. Keep cold.

100 mL   10 mM K_2HPO_4-KH_2PO_4 buffer, pH 7.2. Make the following stocks first:

$\qquad$ 10 mM K_2HPO_4: 0.174 g/100 mL.

$\qquad$ 10 mM KH_2PO_4: 0.136 g/100 mL.

$\qquad$ Mix $\quad$ 72 mL K_2HPO_4 and

$\qquad\qquad\quad$ <u>28 mL KH_2PO_4</u> to get

$\qquad$ 100 mL buffer at pH 7.2. Check pH.

100 mL   Assay medium:

$\qquad$ 0.3 M mannitol: 5.4 g/100 mL.

$\qquad$ 10 mM KCl: 0.075 g/100 mL.

$\qquad$ 5 mM $MgCl_2 \cdot 6H_2O$: 0.1 g/100 mL.

$\qquad$ 0.1% BSA (fatty acid free): 0.1 g/100 mL.

$\qquad$ in 10 mM K_2HPO_3-KH_2PO_4 buffer (from above), pH 7.2. Keep cold.

50 mL   50 mM DNP in ethyl alcohol: 0.09 g/50 mL. Wear latex gloves. NOTE: DNP waste will be present in the aspirator, mixed with other waste materials; label and dispose of properly.

10 mL   15 mM ADP: 0.06 g/10 mL.

100 mL   150 mM succinate: 1.77 g/100 mL. Use succinic acid and adjust pH to 7.2 with NaOH.

To isolate mitochondria: The day before, cut at least 100 g of discs and incubate them in aerated water. Make two batches of mitochondrial preparation (aged and fresh) at the same time. Grind each type of potato separately and filter. Pour each into **labeled** centrifuge tubes and do all the subsequent steps simultaneously.

1) In advance, set up an ice bucket for each batch containing a 400 mL beaker, funnel, two layers cheesecloth and nine 50-mL centrifuge tubes.
2) Grind 150 g of each tissue in 120 mL homogenization mix and filter through the cheesecloth. Divide each extract into four labeled centrifuge tubes and spin at 1500 g for 20 minutes.
3) Transfer the supernatant to labeled centrifuge tubes and sediment at 12,000 g for 20 minutes.
4) Discard the supernatant and resuspend the pellet in 4 mL of homogenization mix and combine the pellets into one tube. Spin at 1500 g for 10 minutes.
5) Transfer the supernatant to another labeled centrifuge tube and sediment at 12,000 g for 20 minutes.
6) Resuspend the pellet in a small volume of washing medium.
7) Determine the protein content per mL. Reagents for protein determinations are available from most chemical companies. Adjust the protein concentration to 1 mg/mL. At least 5 mL of mitochondrial preparation are needed of each for a class of 12 students working in groups of two.

Plant material: Potatoes (*Solanum tuberosum*) may be purchased at the supermarket. Discs may be prepared by slicing the potatoes with a Feemster Famous vegetable slicer or other slicer then cutting through a stack of slices with a #1 (for Part I) or #11 (for mitochondria) cork borer. Age the discs by incubating them in aerated water at room temperature for 24 hours. Fresh discs may be cut just before class and soaked in distilled water. For Part I soak some aged and fresh discs in 10 mL of water containing 0.1 mL of 50 mM KCN. It is important for Part I that the discs be of the same size and thickness. When aging discs for the mitochondrial preparation, less care may be taken in cutting the discs.

Class time: It takes about one-half hour to set up Part I. Part II may be completed during the two hours before the final readings are taken for Part I. If necessary, the fresh and aged treatments for Parts I and II may be divided among the class.

EXPERIMENT 9: Tissue-Water Relations in Potato

Solutions:

2000 mL   Ice-salt bath: 300 g salt/2000 mL tap water. Use store-bought salt and 2000 mL plastic beakers. The salt must be completely dissolved before freezing. Prepare and freeze several days before class.

500 mL each   Sucrose solutions. Store-bought sugar is adequate and cheaper than sucrose purchased from a chemical supply company. Make these solutions just before use or refrigerate to prevent contamination. Warm to room temperature before class.

> 0.1 molal sucrose: 17.1 g + 500 mL water.
> 0.2 molal sucrose: 34.2 g + 500 mL water.
> 0.3 molal sucrose: 51.3 g + 500 mL water.
> 0.4 molal sucrose: 68.4 g + 500 mL water.
> 0.5 molal sucrose: 85.5 g + 500 mL water.
> 0.6 molal sucrose: 102.6 g + 500 mL water.
> 0.7 molal sucrose: 119.7 g + 500 mL water.

Plant material: Potatoes (*Solanum tuberosum*) may be purchased at the supermarket.

Equipment: Thermometers appropriate for freezing point determinations may be obtained from Thomas Scientific. Large culture tubes used as "freezing tubes" may be obtained from Bellco Glass, Vineland, NJ 08360. Balances for Part I must be readable to 0.01 g.

Alternative: Use beet roots instead of potatoes and soak thin slices in the sucrose solutions from Part I. Observe section with a microscope and look for 50% plasmolysis. If balances are not available, measurement of the change in length of the potato cylinders will work just as well.

Class time: About 3 h. One or more short experiments such as Pressure Bomb, Translocation or Mineral Nutrition may be started during the same class.

EXPERIMENT 10: The Pressure Bomb and Determination of Water Potential

Plant material: Cocklebur (*Xanthium strumarium*) plants, 4 weeks old, grown under continuous light. Withhold water from half the plants 6 days before class. Seed may be obtained from Carol Reiss, Section of Plant Biology, Cornell University, Ithaca, NY 14853.

Equipment: Pressure Bomb (Model 3000 series Plant Water Status Console) may be purchased from Soilmaster Equipment Corp., P.O. Box 30025, Santa Barbara, CA 93105.

Class time: About 20 minutes per group. May be completed in conjunction with another experiment such as Tissue-water Relations.

EXPERIMENT 11: Transpiration and the Mechanism of Guard Cell Movement

Solutions:

The first four solutions listed are made at 2X final concentration. Each will be mixed with an equal amount of Ca-MOPS to give the final correct concentration.

500 mL   200 mM KCl: 7.4 g/500 mL.
250 mL   400 mM mannitol: 18.2 g/250 mL.

250 mL	200 mM choline chloride: 7 g/250 mL.
250 mL	200 mM KCl (3.7 g) and 2×10^{-5} M abscisic acid (3 mg) as one 250 mL solution. Boil to dissolve the abscisic acid.
1500 mL	5 mM MOPS: 1.57 g/1500 mL. Adjust the MOPS with CaO: 0.1 g/1500 mL.

Dilute each of the first four solutions by adding an equal amount of the Ca-MOPS solution to get these final concentrations:

1000 mL	100 mM KCl in Ca-MOPS.
500 mL	200 mM mannitol in Ca-MOPS.
500 mL	100 mM choline chloride in Ca-MOPS.
500 mL	100 mM KCl and 10^{-5} M ABA in Ca-MOPS (cover with foil to keep dark).

For the demonstration slides:

20 mL	$Na_3CO(NO_3)_6$ in 10% acetic acid. Add 2 mL concentrated acetic acid to 18 mL H_2O, then add 7 g sodium cobalt nitrate.
20 mL	5% ammonium sulfide (usually comes as a 20-24% solution, so dilute 1 to 4).

Plant material: Cocklebur (*Xanthium strumarium*) plants, 4 weeks old, grown under constant or nearly constant light. Bean plants are also suitable, and almost any small herbaceous plants would work, but *Xanthium* is best suited for the pressue bomb. *Xanthium* seed may be obtained from Carol Reiss, Section of Plant Biology, Cornell University, Ithaca, NY 14853.
Vicia faba (30 Broad Windsor) plants, 6 weeks old.

Equipment: Humidity tents (Fig. G-2) may be constructed on the lab bench using 0.004" plastic sheeting draped over a frame built with Flexangle (or plastic pipe) and taped into place. The plastic sheeting may be left open at opposite sides and a single tent may be used by two student groups. Almost any

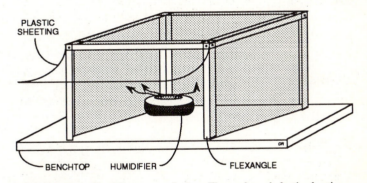

FIGURE G-2. Humidity tent made from Flexangle and plastic sheeting.

type of balance may be used. If torsion balances are selected, the students may counterbalance the plants with flasks containing water.

Alternatives: For Part I: Compare rates in other plants. Soak the roots in ice water. Paint ABA on leaves and determine transpiration rate.

Preparation: Turn on humidifiers at least one hour before class. The *Xanthium* and *Vicia* plants must be placed in the dark for 24 hours before class. To further ensure that the water potential of the *Xanthium* plants is close to zero at the start of the experiment, place the plants in aerated water in the dark several hours before class (this step is not required).

To make the demonstration slides: Place some epidermal strips in 100 mM KCl solution in the light and some in the dark. After one hour, rinse the strips in 0°C distilled H_2O and plunge them into 10 mL of the sodium cobalt nitrate solution at 0°C; keep the strips in this solution for 30 minutes. Rinse the strips in distilled H_2O (at 0°C) until no more color can be removed from them. Then plunge them into 10 mL of ammonium sulfide solution and leave them for two minutes at room temperature. Rinse the strips in distilled H_2O at room temperature; they may be keep in a beaker of distilled H_2O in the refrigerator until needed (or made into permanent slides).

Class time: 3-3.5 h. There is adequate time between measurements to set up other short experiments or make follow-up observations for other experiments. The measurement of water potential with the pressure bomb is the major bottleneck in this experiment. Make sure that students do not cut the petioles if they do not have immediate access to the pressure bomb, as the petioles will dry out. If several groups share a humidity tent, they must try to coordinate the timing of shifts from one treatment to another.

EXPERIMENT 12: Mineral Nutrition

Solutions (It is convenient to make stock solutions in quantity and freeze them for long term storage):

1000 mL	0.5 M $Ca(NO_3)_2 \cdot 4H_2O$: 118 g/1000 mL.
1000 mL	0.5 M KNO_3: 50.5 g/1000 mL.
1000 mL	0.5 M $MgSO_4 \cdot 7H_2O$: 123 g/1000 mL.
1000 mL	0.5 M KH_2PO_4: 68 g/1000 mL.
1000 mL	0.5 M $NaNO_3$: 42.5 g/1000 mL.
1000 mL	0.5 M $MgCl_2 \cdot 6H_2O$: 102 g/1000 mL.
1000 mL	0.5 M Na_2SO_4: 70 g/1000 mL.
1000 mL	0.5 M $NaH_2PO_4 \cdot H_2O$: 69 g/1000 mL.
1000 mL	0.5 M $CaCl_2 \cdot 2H_2O$: 73.5 g/1000 mL.
1000 mL	0.5 M KCl: 37 g/1000 mL.

SUNFLOWER SEEDLINGS

AIR SUPPLY

COVER WITH 3 HOLES

BROWN BOTTLE FILLED WITH HYDROPONIC SOLUTION

1000 mL	FeNaEDTA: Add 7.44 g tetrasodium EDTA to 800 mL distilled water. Once dissolved, add 5.56 g $FeSO_4$ and leave overnight. Filter out the precipitate and bring to 1000 mL.
1000 mL	Micronutrients: one solution containing:

H_3BO_3: 2.88 g.

$MnCl_2 \cdot 4H_2O$: 1.81 g.

$ZnSO_4 \cdot 7H_2O$: 0.22 g.

$CuSO_4 \cdot 5H_2O$: 0.08 g.

H_2MoO_4: 0.02 g.

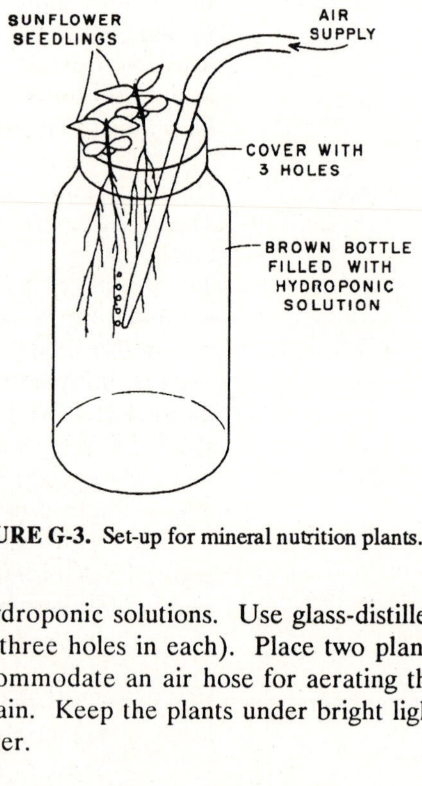

FIGURE G-3. Set-up for mineral nutrition plants.

Follow the schedule on page 108 and make up two of each of the hydroponic solutions. Use glass-distilled water and brown bottles with aluminum weighing pans as lids (punch three holes in each). Place two plants (either corn or sunflower) in each hole and use the third hole to accommodate an air hose for aerating the solution (Fig. G-3). Remove the remaining endosperm of the corn grain. Keep the plants under bright light (14-15 h/day) and maintain the water level with added glass-distilled water.

Plant material: Sunflower (*Helianthus annuus* var. Sungold or other variety) seed is available from Agway. Seedlings should be about 3 weeks old, grown in #4 washed sand.
Corn (*Zea mays* var. Golden Cross Bantum or other variety) seed is available from Agway. Seedlings should be about 3 weeks old, grown in #4 washed sand.

Class time: If this experiment is set up as a demonstration, only about 15 minutes should be needed each week for 3 weeks.

Alternatives: The students may make up the solutions themselves following the chart in the text. The instructor may then code eight of the solutions as unknowns.

EXPERIMENT 13: Nitrate Reductase: The Transformation of *Chlamydomonas*
NOTE: The use of recombinant DNA requires a permit. Check with your college administration for permission to do the following experiment (Section III-C, Fed. Reg. May 87, 1986, p. 16961, NIH Guidelines). A Memorandum of Understanding and Agreement must be submitted at the time the experiment is performed.

Solutions:
Part I.

2000 mL	SGII-NH$_4$ medium:
	    NaH$_2$PO$_4$: 7.34 g.
	    K$_2$HPO$_4$: 2.3 g.
	    Na acetate (NaC$_2$H$_3$O$_2$ • 3H$_2$O): 4 g.
	    20 mL of 5% Na-citrate.
	    20 mL of 0.1% FeCl$_3$ (must be freshly made).
	    20 mL of 0.4% CaCl$_2$ • 2H$_2$O.
	    20 mL of 1.5% MgSO$_4$.
	    20 mL of 3% NH$_4$NO$_3$.
	    20 mL of trace elements stock.
	Divide into appropriate quantities (see Plant Material) and cover with a cotton or foam plug and foil. Autoclave. Reserve 600-700 mL for making agar plates. Additional medium will be required if Part II (optional) is completed.

1000 mL	Trace element stock. One solution containing:
	    H$_3$BO$_3$: 100 mg.
	    ZnSO$_4$ • 7H$_2$O: 100 mg.
	    MnSO$_4$ • 4H$_2$O: 40 mg.
	    CoCl$_2$ • 6H$_2$O: 20 mg.
	    Na$_2$MoO$_4$ • 2H$_2$O: 20 mg.
	    CuSO$_4$: 4 mg.

25 plates	To 700 mL of the SGII-NH$_4$ medium add 10.5 g agar (to make 1.5% agar). Autoclave and allow the medium to cool somewhat before pouring. Pour into sterile petri dishes under sterile conditions; if possible let the dishes cool under sterile conditions before capping to prevent excessive condensation.

1000 mL	2X concentration of SGII-NO$_3$ medium. This medium contains the same components as the recipe for SGII-NH$_4$ medium but uses KNO$_3$ instead of NH$_4$NO$_3$.
	    NaH$_2$PO$_4$: 7.34 g.
	    K$_2$HPO$_4$: 2.3 g.
	    Na acetate (NaC$_2$H$_3$O$_2$ • 3H$_2$O): 4 g.
	    20 mL of 5% Na-citrate.
	    20 mL of 0.1% FeCl$_3$ (must be freshly made).
	    20 mL of 0.4% CaCl$_2$ • 2H$_2$O.
	    20 mL of 1.5% MgSO$_4$.
	    20 mL of 10% KNO$_3$.
	    20 mL of trace elements stock.
	Autoclave. Dilute 250 mL with distilled water to make 500 mL of regular strength SGII-NO$_4$ medium (400 mL for rinsing the agar plates and 100 mL for resuspending cells). Use the remaining 750 mL of 2X concentration medium to make agar plates. Additional medium will be required if Part II is done.

50 plates	SGII-NO$_3$ medium and 1.5% agar. Rinse the agar beforehand to remove ammonium by placing 7.5 g agar in each of three graduated bottles and bringing to 500 mL with distilled water. After the agar has settled, replace the water and shake; repeat once more, then bring to 250 mL with distilled water. Finally, add sufficient 2X SGII-NO$_3$ medium to bring to 500 mL. Autoclave and let the medium cool before pouring, then pour into sterile petri

dishes. Under sterile conditions, rinse the surface of the gelled agar with sterile SGII-NO$_3$ medium.

100 mL   20% (wt/vol) polyethylene glycol (MW = 8000). Autoclave.

200 mL   70% ethyl alcohol.

Part II.

500 mL   SGII-NO$_3$ medium. See Part I instructions.

500 mL   SGII-NH$_4$ medium. See Part I instructions.

200 mL   Nitrate buffer (pH 7.1):

         250 mM NaH$_2$PO$_4$ · H$_2$O: 6.9 g/200 mL.

         50 mM NaNO$_3$: 0.85 g/200 mL.

         50 μM dithiothreitol (DTT): 1 mL of 10 mM DTT (0.15 g/100 mL).

         Check pH and adjust if necessary.

1 mL   Nystatin: 5000 units/mL. Make fresh and store on ice.

400 mL   Sulfanilamide: 4 g/400 mL of a solution of 100 mL concentrated HCl and 300 mL dist. water.

400 mL   N-1-naphthylethylenediamine dihydrochloride (NEED): 0.08 g/400 mL.

400 mL   Nitrite standard solution (25 nmol/mL): Make a 10 mM nitrite stock solution (85 mg KNO$_2$/100 mL). Dilute 1 mL of stock to 400 mL.

Plant material: *Chlamydomonas reinhardtii*, nit1, cw-15 can be obtained from the Plant Science Center, Biotechnology Building, Cornell University, Ithaca, NY 14853. The culture is received on a "slant" of agar and may be kept in that form for several months. Eight to ten days before class, under sterile conditions, transfer a small portion of the *Chlamydomonas* from the slant to 100 mL of sterile SGII-NH$_4$ medium (autoclaved). Gently shake the culture in the light at room temperature for about one week until the cell concentration is about 1-2 x 10^6 cells/mL. The concentration may be determined by adding two drops of 0.25% iodine (in ethyl alcohol) to a 2-mL aliquot of the culture, then counting the cells on a hemacytometer. Under sterile conditions, transfer a sufficient number of cells to 600 mL of sterile SGII-NH$_4$ medium in a 2000-mL flask to make a final concentration of 0.5 x 10^5 cells/mL. Shake the culture at room temperature in the light for one to two days. The final concentration of cells should be between 1 and 2 x 10^6 cells/mL. Either sediment the entire culture (at low speed, 3000 x g or below for 5 minutes) or transfer 5 x 10^7 cells (about 30 mL) to 15-mL polypropylene centrifuge tubes and sediment (in two steps!). Resuspend in one-hundredth the volume of SGII-NO$_3$ medium. If the entire culture has been sedimented and resuspended, students may transfer 0.3 mL of the culture to the polypropylene tubes during class. Sterile conditions must be maintained at all steps. It is recommended that several test runs of cell culture be performed to determine the optimum timing for growth of the cells under the conditions present in your laboratory. It is important that the cells be at the correct stage of growth at the time of transformation; a reduction in the efficiency of transformation will occur if older cells taken from a more concentrated culture are used.

Plasmids (pMN24) containing the nitrate reductase gene may be obtained from the Plant Science Center as well. They may be amplified using standard molecular biology techniques as found in "Molecular Cloning, a Laboratory Manual," 2nd ed., Cold Spring Harbor Laboratory Press, 1989.

For **Part II**, transfer (under sterile conditions) a single colony of transformed cells to 100 mL of sterile SGII-NO$_3$ medium. Grow as before. Final cell concentration should be between 1 and 2 x 10^6 cells/mL. Subculture to new liquid medium of the cells are growing too rapidly. Prepare the nit1 mutant cells as for Part I. Before class, spin down the cells and resuspend in 50 mL of the nitrate buffer. If you wish to determine activity per mg of chlorophyll, you may check the chlorophyll concentration using the method described for Expt. 5 (page 264) and adjust the concentration to 0.2 mg/mL.

Equipment: If a laminar flow hood is unavailable, a sterile area may be created by placing plastic sheeting on supports around a small area of lab bench. If the entire area is wiped with 70% ethyl alcohol and air exchange is prevented, the area will remain sterile. Repeated wipes with alcohol are helpful. UV lamps are not necessary and are dangerous if left on during sterile transfers.

Alternatives: Use plasmids which have been cut with restriction enzymes as well as supercoiled plasmid and compare the efficiency of transformation. After the transformed colonies have grown, test for nitrate reductase.

Cleanup: All contaminated liquid or solid wastes must be decontaminated (by autoclaving) before disposal.

EXPERIMENT 14: Ion Uptake by Potato Tuber Discs

NOTE: The use of radioactive materials requires a permit. Check with your college administration for permission to use ^{32}P.

Solutions:

500 mL	0.02 mM KH_2PO_4 (pH 6.0) containing 0.2 mM $CaSO_4 \cdot 2H_2O$. Make 100 mL of 20 mM KH_2PO_4 (0.27 g/100 mL) as a stock solution. Add 0.5 mL of this to 400 mL distilled H_2O. Make 100 mL of 10 mM $CaSO_4$ (0.17 g/100 mL) and add 10 mL to the above solution. Adjust pH to 6.0 and bring to 500 mL.
100 mL	0.2 mM KH_2PO_4 (pH 6.0) containing 0.2 mM $CaSO_4$. Use the above stock solutions. Add 1 mL 20 mM KH_2PO_4 and 2 mL 10 mM $CaSO_4$ to 90 mL distilled water. Adjust the pH to 6.0 and bring to 100 mL.
2000 mL	1 mM KH_2PO_4 (pH 6.0) RINSE: 0.27 g/2000 mL. Adjust pH.
4000 mL	1 M KH_2PO_4 (pH 6.0) CLEANUP: 544 g/4000 mL. Add 56 g KOH before adjusting volume to adjust pH. Check pH.
50 mL	1 mM CCCP: 0.01 g/50 mL DMSO. Dispose of properly. Radioactive waste containing CCCP must be labeled for **both**.
50 mL	60 mM KCN: 0.195 g/50 mL. Use a face mask while weighing out the cyanide. Dispose of properly. Radioactive waste containing KCN must be labeled for **both**.
10 mL	$KH_2{}^{32}PO_4$: $1.8\,\mu$Ci/mL H_2O. Dispose of properly (see Appendix E).

If Liquid Scintillation Counting is used:

900 mL   Scintillation cocktail. NOTE: Premade, biodegradable scintillation cocktails are now commercially available and are recommended. The following recipe contains toluene and must be used with care. Prepare and dispense in a fume hood.

        600 mL   purified toluene.
        0.09 g   POPOP (dissolve in toluene first).
        4.5 g   PPO.
        300 mL   Triton X-100.

It is convenient to keep the cocktail in a Repipet set for the desired amount (3 mL) and stored in a fume hood. Dispose of properly. Wear toluene-resistant gloves when using either the scintillation fluid described here or the biodegradable types.

Plant material: Potatoes (*Solanum tuberosum*) may be purchased at the supermarket. Prepare discs by slicing the potatoes with a Feemster Famous vegetable slicer or other slicer, then cut through several slices with a #11 cork borer. Age the discs by placing them in aerated water overnight at room temperature. Fresh discs may be cut just before class and soaked in distilled water. It is important that the discs be of the equal thickness and size.

Class time: About 3 h. It is possible to complete both this experiment and Ion Transport in one laboratory class.

Cleanup: The hands and feet of each student should be monitored with a hand-held Geiger-Müller counter before they are permitted to leave the classroom. Lab benches and materials used in the class should be wiped (to test for removable radioactivity, see Appendix E) and monitored as well. For maximum sensitivity, monitor 1 cm from the surface. Collect and dispose of radioactive materials as required by local ordinance. Separate containers may be required for degradable radioactive waste which contains cyanide and/or CCCP.

EXPERIMENT 15:   Ion Transport and Electrochemical Potentials

Solutions:

500 mL	3 M KCl:	112 g/500 mL.
100 mL	100 mM KCl.	Use 3.3 mL of 3 M KCl/100 mL.
100 mL	10 mM KCl.	Use 10 mL of 100 mM KCl/100 mL.
100 mL	1 mM KCl.	Use 1 mL of 100 mM KCl/100 mL.
100 mL	0.1 mM KCl.	Use 0.1 mL of 100 mM KCl/100 mL.
1000 mL	Bathing solution. Make as one 1000 mL solution:	

    0.5 mM NaCl: 0.03 g/1000 mL.

    0.5 mM KCl: 6.25 mL of 100 mM KCl/1000 mL.

Equipment: Use a pH meter (or other millivoltmeter), two calomel electrodes and a 1 mm silver wire with a AgCl coating (soak the silver wire in chlorine bleach for 20 minutes). To zero for concentration determinations, place the AgCl electrode (input) and calomel reference electrode into the same 3 M KCl solution and adjust the meter to 0 mV. Readings will be made on the + mV scale. To reset the zero for the electrical potential difference readings, place the two Calomel electrodes in the same 3 M KCl solution (or connect two 3 M KCl solutions with a salt bridge) and adjust the reading to zero. The electropotential difference may be read on either the + or - mV scale.

To make salt bridges: Add 2 g of agar to 100 mL of 3 M KCl and autoclave to dissolve the agar. Clamp one end of a 1-ft long piece of Tygon tubing and cut a small (2-3 mm) slit in the tubing, near the clamp. Insert the tip of a 15-cm piece of fine Tygon tubing in the slit. Put the other end of the fine tubing in the agar solution and apply suction on the free end of the large Tygon tubing until the agar solution fills the fine tubing. Make as many salt bridges as you will require and store them in 3 M KCl.

Plant material: 4-5 week old cocklebur (*Xanthium strumarium*) plants work best, but bean or other plants are acceptable. One day before class, remove the stem near the base and slip a 1 inch-long piece of Tygon tubing over the top of the stem just above the roots. The fit must be tight, but the stem should not be damaged. Place the plants in beakers containing bathing solution and aerate overnight. Sufficient exudate fluid will accumulate within a day.

Class time: About 15 minutes per group.

EXPERIMENT 16: Translocation of Labeled Sucrose
NOTE: The use of radioactive materials requires a permit. Check with your college administration for permission to use ^{14}C.

Solutions:

500 mL	5 mM dinitrophenol (DNP): 0.45 g/500 mL ethyl alcohol. Dispose of properly.
5 mL	^{14}C-sucrose: 50 μCi/5 mL. Dispose of properly (see Appendix E).
100 mL	20 μM sucrose: 0.02 mL 0.1 molal sucrose (from Expt. 10)/100 mL H_2O.

Plant material: 3-week-old and 6-week-old Red Kidney bean (*Phaseolus vulgaris* var. Redcloud) plants. The older plants should have developing fruits.

Equipment: Kodak X-Omat RP film, developer and fix may be obtained from Standard Medical Equipment, P.O. Box 15840, Baltimore, MD 21263 or Sigma Chemical Co. Use a darkroom safelight with Kodak GBX-2 filter.

Class time: About 15 minutes per group is needed on the first day. Follow up the next day takes about 30 minutes. During the next class each group should spend about 15 minutes in the darkroom.

Cleanup: Wipe tests should be performed at lab benches where radioactive materials have been used. Moist circles of filter paper (2.5 cm) must be wiped over the area (about 100 cm^2), then placed in a 20-mL vial containing 5 mL of scintillation cocktail. Place the filter at the bottom of the vial with the wiped side up. The vials should be counted in the Liquid Scintillation Counter for 1-5 minutes each and should have CPM readings comparable to background levels. Areas which reveal higher counts must be carefully cleaned and wiped again.

EXPERIMENT 17: Plant Tissue Culture

Solutions:

1000 mL	15% chlorine bleach: 150 mL/1000 mL.
1000 mL	70% ethyl alcohol in a plastic squeeze bottle.

To make the media: One container of each medium type is needed per student group with at least 30 mL of the appropriate medium in each; the following instructions will provide enough medium for about 30 containers of each type. NOTE: Several kinds of powdered media are available commercially. They are relatively inexpensive and make preparation of the media considerably easier. They may be substituted for the media that follow; however, the hormone concentrations will be different so the information in Table 18-2 will require correction.

1)   Make the following stock solutions:
        $CoCl_2 \cdot 6H_2O$: 2.5 mg/100 mL.
        KI: 83 mg/100 mL.
        $CuSO_4 \cdot 5H_2O$: 2.5 mg/100 mL.
        $Na_2MoO_4 \cdot 2H_2O$: 25 mg/100 mL.
        Thiamine-HCl: 40 mg/100 mL.
2)   Put 1 L of distilled H_2O in a 2-L flask.
3)   Add 3 mL of each of the stock solutions from #1.
4)   Make the following stock solutions, then add the appropriate amounts to the 2-L flask.

Chemical	Stock Solution	Amount To Add
$CaCl_2 \cdot 2H_2O$	15 g/100 mL	8.7 mL
KNO_3	22.8 g/120	30
KH_2PO_4	17 g/200	6
$MgSO_4 \cdot 7H_2O$	37 g/100	3
NH_4NO_3	19.8 g/180	45
H_3BO_3	620 mg/100	3
$ZnSO_4 \cdot 7H_2O$	860 mg/100	3
$MnSO_4 \cdot H_2O$	2.23 g/100	3

5) Dissolve 83.3 mg $FeSO_4 \cdot 7H_2O$ in some distilled water, then add to the above mixture.

6) Dissolve 112 mg Na_2EDTA in some distilled water, then add to the above mixture.

7) Dissolve 50 mg of nicotinic acid in 100 mL of distilled water, then add 3 mL to the above mixture.

8) Dissolve 50 mg of pyridoxine HCl in 100 mL of distilled water, then add 3 mL to the above mixture.

9) Dissolve 1 g of glycine in 100 mL of distilled water, then add 0.6 mL to the above mixture.

10) Finally, add 0.3 g myo-inositol, 3 g casein hydrolysate and 90 g sucrose.

11) Bring to 1500 mL. Then divide into three 500 mL portions and place each portion in a 2000 mL flask.

12) Make the following stock hormone solutions
#1. Kinetin, 10 mg/100 mL. Boil to dissolve.
#2. auxin (IAA), 40 mg/100 mL. Dissolve the IAA first in a minimal amount of 95% ethyl alcohol, then bring to volume with water.

13) Add kinetin and IAA to each of the flasks according to the following schedule. Label each one clearly.

Label	#1 (Kinetin)	#2 (IAA)
Green	0.2 mL	5 mL
White	2	5
Yellow	5	5

14) Bring the volume up to 975 mL in each of the four flasks and adjust the pH to 5.7-5.8 in each flask. Bring up to 1000 mL.

15) Add 12 g of agar to each flask and swirl.

16) Autoclave the three flasks for 30 minutes. Label the caps on the tissue culture containers with green, yellow and white tape while the media are autoclaving. Allow the media to cool somewhat before pouring.

17) Add 30 mL of the appropriate medium to each color coded vial and cap. If possible allow the media to cool further in a sterile area before capping.

18) Autoclave the containers, then leave to cool on the bench top. Once the containers are cool they may be stored in the refrigerator.

Plant material: Tobacco (*Nicotiana tabacum* var. Maryland Mammoth) plants are suggested, but almost any variety of Tobacco with a large pith may be used. Maryland Mammoth seed is available from Carol Reiss, Section of Plant Biology, Cornell University, Ithaca, NY 14853. Plant the seed in large pots several months in advance; keep under lights for 16 h./day.

Equipment: If a laminar flow hood is unavailable, a sterile area may be created by placing plastic sheeting on supports around a small area of lab bench. If the entire area is wiped with 70% ethyl alcohol and air exchange is prevented, the area will remain sterile. Repeated wipes with alcohol are helpful. UV lamps are not necessary and are dangerous if left on during sterile transfers.

Class time: About 15 minutes per group to set up. Weekly observations should take about 10 minutes.

Alternatives: Additional treatments might include media which omit one or both of the added hormones.

EXPERIMENT 18: Plant Movements and the Differential Growth of Plants

Solutions:

20 mL   1% naphthaleneacetic acid (NAA) in lanolin. Dissolve 0.2 g NAA in a minimal amount of ethyl alcohol (about 2 mL). Melt lanolin in a beaker to obtain exactly 20 mL and add the NAA-ethyl alcohol mix to the slightly cooled lanolin. Save extra for Expt. 20.

20 mL   1% triiodobenzoic acid (TIBA) in lanolin. Dissolve TIBA in a minimal amount (2 mL) of dimethylsulfoxide (DMSO). Melt lanolin in a beaker to get 20 mL and add the TIBA-DMSO mix to the slightly cooled lanolin and stir.

Plant material: Sunflower (*Helianthus annuus* var. Sungold) plants, 5 weeks old

Scarlet Runner bean (*Phaseolus coccineus*) plants, 6 weeks old. When the plants are 2-3 weeks old, stake the plants, then wrap and tie half of them clockwise and half counterclockwise.

Pea (*Pisum sativum* var. Progress #9) plants, 4 weeks old.

Oat (*Avena sativa* var. Porter) coleoptiles, grown in the dark, 6 days.

Red Kidney bean (*Phaseolus vulgaris* var. Redcloud) plants, 3-4 weeks old.

Oxalis spp. plants, several weeks old.

Venus flytrap (*Dionaea muscipula*) plants may be obtained from many nurseries. Make sure the traps are large and healthy. They are best grown on peat moss in bright sunlight. Change the peat every 3 months and do not fertilize.

Mimosa pudica plants.

Equipment: A sheet of black construction paper may be cut and folded to provide sufficient cover for the oat coleoptiles (Fig. G-4). A flood lamp (150 W), if placed several feet away on the uncovered side, will provide sufficient light to elicit a phototropic response even in a normally lit room.

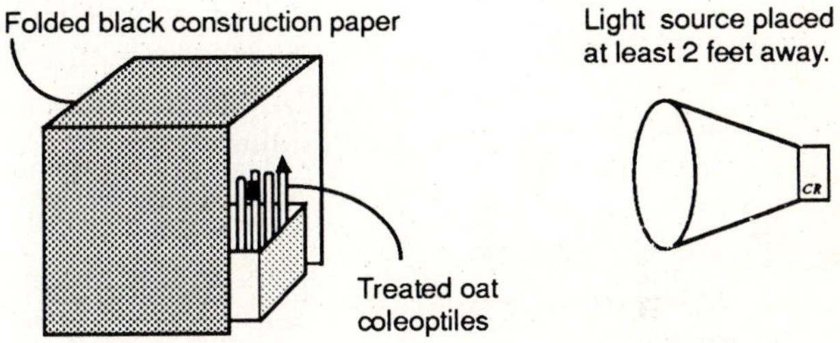

FIGURE G-4. Setup for phototropism experiment.

Monochromatic light is more difficult to provide, since you must be sure that no white light leaks into the box. Wooden boxes may be constructed with a slit on one side, which may be covered with a color filter.

A recorder and silver wire (wired through a junction box to insulated copper wire adapters) are needed to record action potentials (Fig G-5).

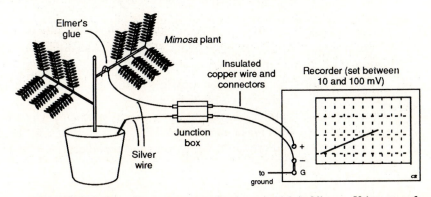

FIGURE G-5. Setup for measurement of action potentials in *Mimosa*. Using a transfer pipet, drip ice water on the leaflets to induce an action potential and cause the petiole to drop.

Preparation: A bean and *Oxalis* plant should be placed in a growth chamber one week before class. Set the light period to run from 8 p.m. to 8 a.m. Another bean and *Oxalis* should be kept in room light.

Class time: 3 h. There is sufficient time for making observations of ongoing experiments.

Alternative: Set up the experiment as a demonstration and combine it with another experiment.

EXPERIMENT 19: **Calcium and Signal Transduction: Cytoplasmic Streaming in *Chara* Cells**

Solutions:

100 mL	10 mM EGTA: 0.38 g/100 mL.
100 mL	10 mM HEPES buffer: 0.24 g/100 mL.
100 mL	10 mM NaCl: 0.06 g/100 mL.
100 mL	10 mM KCl: 0.07 g/100 mL.
100 mL	10 mM $CaCl_2 \cdot 2H_2O$: 0.147 g/100 mL.

100 mL 10^{-4} M $CaCl_2$, buffered with 1 mM EGTA and containing 2 mM HEPES, 0.1 mM NaCl and 0.1 mM KCl. Combine 10 mL of 10 mM EGTA, 20 mL of 10 mM HEPES buffer and add 10.98 mL of 10 mM $CaCl_2$. Add 1 mL each of 10 mM NaCl and 10 mM KCl. Bring almost to volume and adjust the pH to 7.2. Bring to volume. The EGTA will bind Ca^{2+}; the amount of Ca^{2+} to be added has been adjusted to maintain the **free** concentration of Ca^{2+} at 10^{-4} M.

100 mL 10^{-7} M $CaCl_2$, buffered with 1 mM EGTA and containing 2 mM HEPES, 0.1 mM NaCl and 0.1 mM KCl. Combine 10 mL of 10 mM EGTA, 20 mL of HEPES buffer and add 3.99 mL of 10 mM $CaCl_2$. Add 1 mL each of 10 mM NaCl and 10 mM KCl. Bring almost to volume and adjust the pH to 7.2. Bring to volume. If available, use calcium-free distilled water.

Plant Material: *Chara* plants, cut into individual internodal cells and floated on pCa 4 solution an hour before class. *Chara* plants may be obtained from Carolina Biological Supply Company. *Nitella* will work just as well.

Equipment: Build the reaction chambers from glass slides as follows: Use super glue to attach two slides along the edges of the upper surface of a third slide, leaving a 1-cm channel running down the center. Two more slides may be glued to the bottom of the upper slides to stabilize the chamber and keep it from tilting. Glue two 5- to 6-inch pieces of fine silver or copper wire to the chamber. About an inch of wire should be glued along each end of the channel and the wires should protrude from opposite ends of the channel. Use silicone

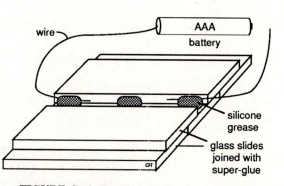

FIGURE G-6. Reaction chamber for inducing an action potential in a *Chara* cell.

grease (which is preferable to, but more expensive than, petroleum jelly) to seal each end of the channel, then make an additional barrier with grease at the center. The completed reaction cell should contain two grease-bound chambers, each containing an attached piece of wire (Fig. G-6).

Class time:	May be completed within one half an hour; may be combined with another experiment such as Plant Movements, GA Bioassay, or Ethylene Production.

EXPERIMENT 20: Apical Dominance

Solutions:

20 mL	1% naphthaleneacetic acid (NAA) in lanolin. (See Expt. 18.)
100 mL	1.5 mM 6-benzylaminopurine (BA) + 0.1% Tween-20: 0.034 g BA/100 mL. Boil to dissolve. Add 0.1 mL Tween-20 last. Do not refrigerate.

Plant material:	Red Kidney bean (*Phaseolus vulgaris* var. Redcloud) plants, 3 weeks old, 2/pot.

Class time:	About 30 minutes set up time. Follow up with observations twice a week for 2 weeks. May be started in conjunction with the Senescence and Leaf Abscission experiments.

EXPERIMENT 21: Leaf Senescence

Solutions:

100 mL	0.2 mM 6-benzylaminopurine (BA): 4.5 mg/100 mL. Adjust to pH 6. Boil to dissolve.
2000 mL	0.1 mM kinetin: 0.043 g/2000 mL. Adjust to pH 6. Boil to dissolve.
500 mL	1 mM spermidine (trihydrochloride): 0.127 g/500 mL. Adjust to pH 6-7.

Plant material:	Red Kidney bean (*Phaseolus vulgaris* var. Redcloud) plants, 3 weeks old, 1/pot. Do not fertilize. Wheat (*Triticum aestivum* var. Frankenmuth) seedlings, grown in vermiculite, 6 days old. Students may use any other plants available for Part III. *Xanthium* plants will work well.

Equipment:	Munsell Color Charts for Plant Tissue (2.5 GY and 5 GY) are available from Macbeth Division of Kollmorgen Instruments, P.O. Box 230, Little Britain Rd., Newburgh, NY 12551-0230. As an alternative, paint chips may be used to make color charts.

Class time:	Can be set up in 30 minutes. Observations must be made every few days for two weeks.

EXPERIMENT 22: Hormones and Leaf Abscission

Solutions:
PART I.

20 mL	1% naphthaleneacetic acid (NAA) in lanolin. (See Expt. 18.)

PART II.

50 dishes	3% agar (make 2 liters): 60 g/2000 mL. Autoclave. Pour about 35 mL/plate into sterile plastic petri dishes. Let cool then store in the refrigerator.
50 mL	$1 \mu g/mL$ (1 ppm) cycloheximide (CHI) in lanolin. Dissolve 1 mg CHI in 10 mL 95% ethyl alcohol. Melt lanolin in beaker to obtain 50 mL. Add 0.5 mL of the CHI-ethyl alcohol mix to slightly cooled lanolin and mix.
100 mL	0.2 mL/L (200 ppm) ethephon: 0.1 mL Ethrel/100 mL or 0.5 mL Florel/100 mL. Ethrel may be obtained from Union Carbide, P.O. Box 12014, T.W. Alexander Dr., Research Triangle Park, NC 27709. Florel is available from AmChem, Agricultural Chemicals Div., Ambler, PA 19002.
100 mL	0.05 M phosphate buffer, pH 7.5:

Make 0.05 M KH_2PO_4: 0.34 g/50 mL.
and 0.05 M K_2HPO_4: 0.87 g/100 mL.
for pH 7.5 use:
16 mL 0.05 M KH_2PO_4 and
<u>84 mL 0.05 M K_2HPO_4</u> to make
100 mL buffer at pH 7.5. Check pH.

PART III.

1000 mL   12 mM HCl: Bring 100 mL of HCl to volume with distilled water.

100 mL   0.05% CPTS stain in 12 mM HCl: 5 g of copper phthalocyanine tetrasulfonic acid (tetra sodium salt) in 100 mL of 12 mM HCl (from above).

Plant material: Red Kidney bean (*Phaseolus vulgaris* var. Redcloud) plants, 4 weeks old, 2/pot. *Albizzia julibrissin* plants or other plants with compound leaves.

Alternatives: Part II: Almost any compound leaf may be used. Students may collect plant material themselves (if you live in an area of suitable climate). A more obvious abscission zone is present at the base of each pinna, so if plant material is abundant, explants may be made from sections of rachis and pinna.

Class time: Set up time is about 1 hour.

EXPERIMENT 23: Ethylene Production and Flower Senescence

Solutions:

1500 mL   5 mM KCl: 0.56 g/1500 mL. Adjust to pH 6.

500 mL   1 mM 6-benzylaminopurine (BA) in 5 mM KCl: 0.113 g BA/500 mL KCl. Boil to dissolve in 500 mL of above solution. Adjust pH to 6. Replace any water lost in boiling.

500 mL   10 μM 1-aminocyclopropane carboxylic acid (ACC). Dissolve 5 mg ACC in 10 mL of distilled water. Transfer 1 mL of this solution to 490 mL of the KCl solution above. Adjust to pH 6 and bring to 500 mL. Keep dark and refrigerate.

Plant material: Morning Glory seed (*Ipomoea tricolor* var. Heavenly Blue) may be obtained from most seed companies. Plant about 12 weeks before class and supplement the day length to 14 hours during the winter. This experiment will only work if class is scheduled for the afternoon. At about 10 a.m., cut the flowers into rib sections and float, rib side down, in petri dishes containing the above solutions.

Standards: Canisters of ethylene mixed with air are available from Airco and Matheson Gas Products.

Alternatives: Use Na benzoate, high auxin, or $AgNO_3$ in 5 mM KCl as the incubation solutions. If access to the GC is limited, a single reading may be taken after several hours to determine the amount of ethylene accumulated in each treatment. As another control, include a vial containing only blue petal parts (which produce little C_2H_4) to emphasize the importance of selecting the most appropriate tissue.

Class time: Measurements are made once an hour (e.g., at 2, 3 and 4 p.m.), allowing this experiment to be completed in conjunction with another experiment such as the GA bioassay.

EXPERIMENT 24: Bioassay for Gibberellins

Solutions:

800 mL   Gibberellic acid (GA), 10^{-6} g/0.2 mL: 4 mg/800 mL. Boil to dissolve and bring back to volume. Use for the solutions below (and for the unknowns, see Teacher's Manual).

100 mL   GA: 10^{-7} g/0.2 mL. Take 10 mL of the first solution and bring to 100 mL.

50 mL   GA: 10^{-8} g/0.2 mL. Take 0.5 mL of the first solution and bring to 50 mL.

50 mL   GA: 10^{-9} g/0.2 mL. Take 0.05 mL of the first solution and bring to 50 mL.

See Teacher's Manual for suggested unknowns.

Plant material: Lettuce seeds (*Lactuca sativa* var. Buttercrunch) may be purchased at Agway or other seed company. Allow the seeds to equilibrate for a week by placing the seed in a beaker, which is placed in larger beaker containing a few mL of water. About 24-26 hours before class, germinate the seed on three layers of No. 1 filter paper in a sealed petri dish. It is convenient to line a plastic tray with a layer of hard styrofoam ("architectural foam") sheeting in which holes are cut to accommodate the shell vials (Fig G-7). Use a #15 cork borer. Each vial containing the seedlings for the bioassay should receive 0.1 mL distilled water one day and three days after class.

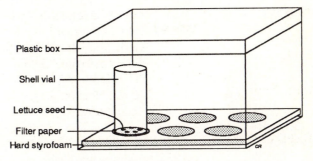

FIGURE G-7. Box for lettuce seed bioassay.

Class time: Set up takes between 20 and 30 minutes. The follow up after four days will require about 30 minutes.

EXPERIMENT 25: α-Amylase: Location and Timing in Wheat Seed Germination

Solutions:

Part I.

500 mL   10 mM citric acid-sodium citrate buffer at pH 5. Make 500 mL of 10 mM sodium citrate (1.47 g/500 mL) and 500 mL of 10 mM citric acid (1.05 g/500 mL). Combine:

         325 mL 10 mM sodium citrate and

         <u>+175 mL 10 mM citric acid</u> to make

         500 mL 10 mM citric acid-sodium citrate buffer at pH 5.0. Check and adjust pH.

     Keep refrigerated.

500 mL 10^{-6} M gibberellic acid (GA) in 0.01 M HEPES-EGTA-Ca^{2+} buffer at pCa 4 and pH 6.

         HEPES: 1.19 g.

         EGTA: 1.9 g.

         $CaCl_2 \cdot 2H_2O$: 0.54 g.

     Adjust to pH 6. Add 1.75 mg of GA. Boil to dissolve. Check volume.

Part II.

100 mL   10% SDS stock: 10 g/100 mL. Do not refrigerate.

50 mL   0.5 M Tris-HCl stock: 3 g/50 mL. Adjust to pH 6.8 with HCl.

100 mL   1.5 M Tris-HCl stock, pH 8.8: 18.2 g/100 mL. Adjust to pH 8.8 with HCl.

1000 mL   Tank buffer: 0.025 M Tris (pH 8.3), 0.192 glycine, 0.1% SDS. Combine:

Tris: 3 g.

glycine: 14.4 g.

10 mL of 10% SDS.

Bring to 900 mL. Adjust pH and bring to 1000 mL with distilled water.

Wear latex gloves while preparing the following solutions. The instructions for preparation of these solutions have been adapted from the Hoefer Scientific Instruments manual.

10 mL   2X Treatment buffer: 0.125 M Tris-HCl (pH 6.8), 4% SDS, 20% glycerol, 10% 2-mercaptoethanol. Combine:

2.5 mL of 0.5 M Tris stock.

4 mL of 10% SDS stock.

2 mL of glycerol.

1 mL of 2-mercaptoethanol.

Add a pinch of Bromophenol Blue and bring to 10 mL. Divide into 250-μL aliquots and freeze.

1000 mL   Stain: 0.298% Coomassie Blue R, 0.197% Coomassie Blue G, 20% ethyl alcohol, 7% acetic acid. Combine:

Coomassie Blue R-250: 2.98 g.

Coomassie Blue G-250: 1.97 g.

200 mL of ethyl alcohol.

70 mL of acetic acid.

Bring to 1000 mL with distilled water.

1000 mL   7% acetic acid (destain): 70 mL/1000 mL.

100 mL   acrylamide stock solution. Acrylamide is a neurotoxin and may be absorbed through the skin; wear gloves when working with this compound.

acrylamide: 29.2 g.

Bis: 0.8 g.

Bring to 100 mL with distilled water. Store in the dark at 4°C.

5 mL   10% ammonium persulfate: 0.5 g/5 mL. Freeze in 1-mL lots.

Part III.

1000 mL   1 N HCl: 83 mL concentrated HCl made up to 1000 mL (add acid to water!).

500 mL   Iodine solution:

KI: 2.5 g.

KIO$_3$: 0.18 g.

in 500 mL 2 mM NaOH (0.04 g/500 mL).

500 mL   0.05% starch solution in 0.05 M citric acid-sodium citrate buffer at pH 5.0. Make stocks:

500 mL of 0.05 M sodium citrate: 7.35 g/500 mL.

500 mL of 0.05 M citric acid: 5.25 g/500 mL.

Combine:

292.5 mL 0.05 M sodium citrate   and

+<u>157.5 mL 0.05 M citric acid</u>   to make

450 mL 0.05 M buffer at pH 5.0. Check pH.

BOIL 0.25 g soluble starch/50 mL water. Add this 50 mL to the 450 mL buffer to get 500 mL of 0.05% starch at the proper pH. Do not refrigerate.

Plant material: Wheat seeds (*Triticum aestivum* var. Frankenmuth, Houser or other) are available from Agway. Germinate several hundred seeds in trays between layers of wet paper towels and cover with foil. Germinate seed 24, 48 and 72 h. before class. In addition, 72 h. before class, germinate 200 seeds at 4°C. Forty-eight hours before class, cut these seeds in half and place 100 halves on three layers of filter paper in each of four petri dishes: embryo half, embryo half +GA (see Part I solutions), endosperm half and endosperm half +GA. Use 10 mL of liquid in each and seal; place in the dark.

Preparation: Before class begins, test the enzyme for activity by making extracts of 48-h and 72-h seed. Follow the directions for Expt. 2, Part I and do a time course to determine the reaction time to be used in Part III of this experiment. Choose a time which gives about 85-95% starch lost for the 72-h. extract. The reaction time will be longer for the chilled seeds (see suggested time in text).

To make polyacrylamide gels: Gels are most easily poured using a casting chamber which enables several gels to be poured at once (available from Hoefer Scientific Instruments). Follow the manufacturer's instructions for making 1.5 mm thick gels (the instructions below are adapted from the Hoefer Scientific Instruments instruction manual). When the chamber is ready, **put on latex gloves** and prepare the following solution for the resolving gel (for 4 gels):

13.32 mL of acrylamide stock solution.

10 mL of 1.5 M Tris-HCl, pH 8.8.

0.4 mL of 10% SDS.

16 mL of water.

De-aerate the solution then add the following:

200 μL of 10% ammonium persulfate.

20 μL of TEMED.

Immediately pour the gels to a level about 1 cm below the notch in the white plate, making sure that the solution is at the same level in all the gels. Over-lay each gel with about 1 mL of water saturated isobutanol. Allow the gels to polymerize for about an hour. Rinse off the isobutanol layer with distilled water and invert to drain. Prepare the stacking gel solution:

1.34 mL of acrylamide stock.

2.5 mL of 0.5 M Tris-HCl, pH 6.8.

100 μL of 10% SDS.

6 mL of distilled water.

De-aerate the solution then add the following:

50 μL of ammonium persulfate.

4 μL of TEMED.

With a Pasteur pipet, add this solution to the first gel until the level is about even with the notch in the white plate. Add a comparable amount of solution to each gel. Gently insert a comb into the stacking gel solution of each gel. Allow the stacking gel to set. The gels may be removed from the casting box and individually wrapped in plastic wrap. They will keep in the refrigerator for approximately 3 weeks.

Equipment: Mini-gel operating systems are available commercially at relatively modest cost. Most power supplies will run several gels at the same time if the current required for each is multiplied by the number of gels. Follow the manufacturer's instructions. The small gels described here may be run at about 18 mA for one, or 36 mA for two. Higher current settings will decrease the running time.

Standards: *Bacillus subtilis* α-amylase may be used as a standard, but must be boiled before use as it is a dimer. Wheat α-amylase is about 46 kD. In an extra lane, run prestained molecular weight standards or use a β-amylase standard as well.

Alternatives: Compare wheat to rice, which produces α-amylase in the scutellum.

Class time: 3.5 h.

EXPERIMENT 26: Phytochrome Control of Leaflet Movement in *Albizzia*

Equipment: Red light source (approximately 5 μE m^{-2} s^{-1}): Fluorescent light filtered through red Plexiglas (Fig. G-8).
Far-red light source (approximately 5 μE m^{-2} s^{-1}): Bright incandescent light filtered through pan of water (heat filter) and far-red Plexiglas (Fig. G-8). Far-red Plexiglas is available from Carolina Biological Supply or in large sheets (2 x 4 ft) from Westlake Plastics Company, P.O. Box 127, Lenni, PA 19052.
Darkroom with green safelight.

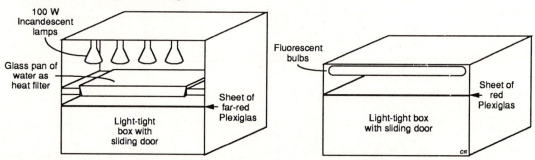

FIGURE G-8. Light boxes for red and far-red light.

Plant material: *Albizzia julibrissin* plants. Adjust the photoperiod of the plants starting about one week before class. The experiment will work best if the light treatments are done early in the photoperiod, so keep the plants in the dark until an hour or so before class.

Class time: 1.5 h.

EXPERIMENT 27: Seed Germination: Light and Hormones

Solutions:
1500 mL   10 mM phosphate buffer. Make stocks for buffer:
                         10 mM KH_2PO_4: 4.08 g/3000 mL.
                         10 mM K_2HPO_4: 0.87 g/500 mL.
                 Then combine:
                         1315.5 mL   10 mM KH_2PO_4 and
                         +184.5 mL   10 mM K_2HPO_4 to make
                         1500 mL   10 mM buffer at pH 6. Check pH.
500 mL   600 μM gibberellic acid: 0.105 g/500 mL phosphate buffer. Boil to dissolve.
500 mL   120 μM abscisic acid (ABA): 16 mg/500 mL phosphate buffer. Boil to dissolve.
500 mL   120 μM kinetin: 13 mg/500 mL phosphate buffer. Boil to dissolve.

Plant material: Lettuce (*Lactuca sativa* var. Grand Rapids) or other light sensitive seed. Imbibe the seed for Part I three hours before the start of class. Place seed on three layers of wet filter in petri dishes and incubate in the dark. Light sensitive lettuce seed is available from Ward's Natural Science, 5100 West Henrietta Road, P.O. Box 92912, Rochester, NY 14692-9012. Test the seed in advance to make sure it is light sensitive. Several days at 30°C should restore lost sensitivity. As an alternative, try germinating the seed in the light and the dark in a series of concentrations of mannitol (0-0.5 M); determine at what concentration the seeds will germinate in the light but not in the dark, then use that concentration of mannitol to imbibe the seed (to be consistent, the hormone solutions should also contain the same concentration of mannitol).

Equipment:	Red light source (approximately 5 μE m^{-2} s^{-1}): Fluorescent light filtered through red Plexiglas (Fig. G-8).
	Far-red light source (approximately 5 μE m^{-2} s^{-1}): Bright incandescent light filtered through pan of water (as heat filter) and far-red Plexiglas (Fig. G-8). Far-red Plexiglas is available from Carolina Biological Supply or in large sheets (2 x 4 ft) from Westlake Plastics Company, P.O. Box 127, Lenni, PA 19052.
	Darkroom with green safelight.
Alternatives:	The students may germinate seed in a series of concentrations of mannitol (see above) if you wish to emphasize the importance of water potential in germination.
Class time:	Set-up takes about 30 minutes for each part and the follow-up after 3 days will take 30-60 minutes.

EXPERIMENT 28: De-etiolation and Phytochrome

Equipment:	Red light source (approximately 5 μE m^{-2} s^{-1}): Fluorescent light filtered through red Plexiglas (Fig. G-8).
	Far-red light source (approximately 5 μE m^{-2} s^{-1}): Bright incandescent light filtered through a pan of water (as heat filter) and far-red Plexiglas (Fig. G-8). Far-red Plexiglas is available from Carolina Biological Supply or in large sheets (2 x 4 ft) from Westlake Plastics Company, P.O. Box 127, Lenni, PA 19052.
	Darkroom with green safelight.
Plant material:	Red Kidney bean (*Phaseolus vulgaris* var. Redcloud) seedlings, grown in the dark, 7 days.
	Pea (*Pisum sativum* var. Progress #9) seedlings, grown in the dark, 7 days.
Class time:	Less than one hour to set up. Students must return outside of class to make observations. Measurement of chlorophyll concentration is not recommended for the follow-up assessment of de-etiolation because the suggested light treatments will not stimulate chlorophyll synthesis sufficiently (see Expt. 29).

EXPERIMENT 29: The Greening of Cucumber Cotyledons

Solutions:

500 mL	20 mM HEPES buffer, pH 7: 2.4 g/500 mL. Use some of this buffer to make the solutions listed below.
50 mL	6-benzylaminopurine (BA): 1 mg/50 mL buffer. Boil to dissolve.
50 mL	δ-aminolevulinic acid (ALA): 0.15 g/50 mL HEPES buffer.
50 mL	cycloheximide (CHI): 2 mg/50 mL HEPES buffer.
500 mL	90% acetone containing 10 mM NH_4OH: add 0.28 mL of concentrated ammonium hydroxide to 500 mL 90% acetone.

Plant material:	Cucumber (*Cucumis sativus* var. Marketmore) seeds, available from Agway and other seed suppliers. Seedlings are grown in vermiculite in the dark at 26°C for 6 days.
Class time:	This experiment should take about one half hour of class time per group. Follow-up on the next day requires precise timing of the 2-hour light exposure. Extraction and measurement of chlorophyll should take about 45 minutes.

GLOSSARY

absorbance	Optical density, which is a measure of the amount of light absorbed by a compound. $A = \epsilon C l$
active metabolism	Metabolism is the sum of the chemical changes in living cells which provide energy for vital processes and activities. Active metabolism implies a large production and turnover of ATP.
active transport	The movement of a compound across a membrane against an electrochemical gradient at the expense of metabolic energy.
aerated	Supplied with air.
agar	A gelatinous, colloidal extract of a red alga used in culture media as a gelling agent.
aliquot	A fractional portion or sample.
apical dominance	The inability of lateral buds to grow when the apex is intact.
autoradiography	The development of an image on photographic film by the radiations from a radioactive substance in close contact with the photographic emulsion.
auxin	Indoleacetic acid, a plant growth regulator.
axillary bud	A bud in the axis of a leaf; a lateral bud. See Appendix F.
azo dye	A dye containing an "azo" group (-N=N-) united at both ends to carbon.
benzyladenine	A cytokinin (a plant growth regulator).
bioassay	A test for the activity of an unknown compound by a comparison of its effect on a test organism with that of a standard compound.
calomel	Mercurous chloride (Hg_2Cl_2).
carborundum	An abrasive compound.
carrier	A transmembrane protein which binds a specific substance and transports it to the other side of the membrane where it is released.
channel	A transmembrane protein which serves as a selective pore in the membrane.
chelator	A compound which forms a soluble complex with a metallic ion.
chromatography	The separation of compounds depending on the compounds affinities for a mobile or stationary phase.
circadian	Approximately a day.
clinostat	An instrument which slowly rotates perpendicular to the earth's gravitational pull, so that plant materials held within do not perceive any directional gravitational stimulus.
coleoptile	The first leaf of a monocotyledonous plant; it forms a protective sheath about the primary bud.
colorimeter	An instrument used for chemical analysis by measurement of a liquid's color. See Appendix C.
compound leaf	A leaf blade in which the leaf is divided into two or more leaflets on a common axis. See Appendix F.
cotyledon	The first leaf produced by an embryo of a seed plant. See Appendix F.
cycloheximide	A protein synthesis inhibitor on 80S (cytosolic) ribosomes.
cytokinins	A group of plant growth substances which includes kinetin and benzyladenine.
de novo	Anew.
dinitrophenol	An uncoupler.

diode array	A bank of photodiodes arranged to simultaneously detect light over a range of wavelengths.
dormancy	In seeds, the inability of a seed to germinate in the presence of adequate water, O_2 and warm temperatures.

electrical potential difference	The difference in electrical charge between two solutions. May refer to the membrane potential.
electrochemical gradient	A difference in electrochemical potential across a membrane or between two cells.
electrode	A conductor used to establish contact with a nonmetallic part of a circuit.
enol group	A hydroxyl group adjacent to a double bond.
epidermal strip	A portion of epidermis and cuticle removed from the surface of a leaf (or other plant part).
ethephon	2-chloroethyl-phosphonic acid; a compound which releases ethylene at high pH.
ethylene	C_2H_4; a plant growth substance which exists as a gas.
explant	A portion of plant material removed for study.

feedback inhibition	The negative effect of a product of a series of reactions on an early step in the series.
free space	The cell walls and intercellular spaces between cells; the apoplast.

gas chromatography	Separation of gaseous (at the chromatograph temperature) compounds dependent on their affinity for a solid versus a gaseous phase.
Geiger-Müller counting	The detection of radioactive emissions by ionization of a contained gas. Electrons, resulting from collision of the emitted particle with the gas are attracted to and affect the electrical properties of a positively charged electrode. The resulting electrical pulses are counted by the electronic system.
germination	The growth or sprouting of seeds.
gravitropism	Growth toward or away from the gravitational pull of the earth in response to the pull of gravity.
guard cells	The two crescent-shaped cells that border and open and close the stomata.

homogenate	A mixture that results from grinding or blending tissue.
hypocotyl	The part of the shoot axis of a seedling below the cotyledons (see Appendix F).

imbibition	The taking up of fluid, resulting in swelling. Seeds must imbibe water before germination can begin.
in situ	In the natural or original position.

Klett unit	The unit of the Klett-Summerson colorimeter. The absorbance (optical density) may be approximated by multiplying by 0.002.

lateral bud	A bud in the axis of a leaf; an axillary bud. See Appendix F.
liquid scintillation counting	The detection and recording of flashes (scintillations) produced when radioactive emissions come in contact with ionizable fluorescing compounds.

mass flow	The movement of groups of molecules along a pressure gradient.

matric potential	Ψ_m, the contribution of water-binding colloids to water potential.
membrane potential	The difference in electrical charge across a membrane.
millivoltmeter	An instrument for measuring in millivolts the differences in potential between different points of an electrical circuit. Recorders and pH meters are both millivoltmeters.
nitrile gloves	Safety gloves resistant to many organic solvents.
node	The point on a stem at which the leaves are attached.
nuclear genome	The complement of DNA found in the nucleus of an organism.
nutation	The spiral movement of a growing plant part.
nyctinasty	"Sleep movements." The folding of the leaves of some plants at night.
osmotic potential	Ψ_s, the contribution made by dissolved solutes to water potential.
passive transport	The movement of compounds across a membrane down an electrochemical gradient.
phloem	The plant tissue, containing sieve elements and companion cells, in which sugars are transported throughout the plant.
photocell	The light detector in the Klett-Summerson colorimeter.
photomultiplier	A light detector of the type found in Liquid Scintillation Spectrometers.
phototropism	The directed growth of plants in response to light.
phototube	The light detector used in the Spectronic 20.
pinnule	One of the ultimate divisions of a double compound leaf.
planchet	A small metal disc with a raised edge used to hold radioactive samples for Geiger-Müller counting.
plasmid	Naturally occurring circular DNA molecule.
polar movement	Unidirectional movement of a substance (such as auxin) from the apex to the base of the plant.
polarizing voltage	The bias voltage given to an electrode, making it positive or negative.
polarography	A method of quantitative analysis based on current/voltage.
polyamines	Plant growth substances; they are involved in, among other responses, the control of senescence.
pressure bomb	An instrument used to measure water potential.
pressure potential	Ψ_p, the contribution made by pressure to water potential.
primary leaf	The first true leaf (see Appendix F).
psychrometer	A hygrometer (consisting of two thermometers) which measures relative humidity. A comparison of the temperature that results from the cooling from evaporation of a wet bulb with the temperature of a dry bulb, constituting a measure of the dryness of the air.
pulvinule	A swelling at the base of a petiolule which functions in turgor-controlled movements.
rachilla	An extension of the petiole of a compound leaf that bears the leaflets.
restriction enzyme	An enzyme which will cut a DNA molecule at a specific site determined by the nucleotide sequence.
significant figures	The figures of a number that end with the last figure to the right that is considered to be correct.
simple leaf	An undivided leaf blade. See Appendix F.

single point mutation   A change in a single base in the DNA sequence of a gene.

solubility   The amount of a substance that will dissolve in another substance.

specific activity   The enzyme activity per mg of total protein.

spectrophotometer   An instrument which measure the absorbance of a substance at a particular wavelength. See Appendix C.

spermidine   A polyamine.

standard curve   A plot of a measurement of some property against a series of known value (e.g. concentration, volume, etc.), which can be used to determine the value of an unknown sample by comparison.

stoma/stomata   An opening or pore in the epidermis of a leaf or plant part, usually surrounded by two guard cells.

supernatant   The solution above the pellet after sedimentation in a centrifuge.

synthetic pathway   A series of biochemical reactions leading to the synthesis of some compound.

tetrapyrrole   A compound containing four pyrrole rings (C_4H_5N).

thigmonasty   A movement in response to touch, where the direction of the movement depends on the plant structure.

thigmotropism   A growth response to touch, where the direction of growth is dependent on the direction of the signal (touch).

thin layer chromatography   Separation of compounds dependent on their affinity for a thin layer of solid versus a liquid phase.

titration   The determination of the concentration of a substance in a solution by measurement of the smallest amount of a reagent needed to react with the substance.

transformation   The incorporation and expression of new DNA into an organism.

translocation   The movement of sugars through the sieve tubes of the phloem in a plant.

transpiration   The loss of water as a vapor from the aerial parts of a plant.

turgor pressure Ψ_p, the contribution made by pressure to the water potential.

uncoupler   A compound which prevents the formation of ATP by chemiosmosis (through the dissipation of the H^+ gradient in chloroplasts or the electrical potential gradient in mitochondria), while allowing electron transport to continue. Some examples of uncouplers are DNP, CCCP and NH_4Cl.

water potential Ψ, an expression of the free energy status of water. A measure of the driving force for water to move.

xylem   The plant tissue, containing xylem elements (dead at maturity), in which water is transported in a transpiration stream.

ABBREVIATIONS USED IN THIS TEXT

A	absorbance
ABA	abscisic acid
ACC	aminocyclopropane carboxylic acid
ADP	adenosine diphosphate
ALA	δ-aminolevulinic acid
ATP	adenosine triphosphate
ATPase	protein which catalyzes the reaction: ATP ADP + Pi
BA	N_6-benzyladenine, a cytokinin
BSA	bovine serum albumin
CCCP	carbonyl cyanide m-chlorophenyl hydrazone, an uncoupler
CHI	cycloheximide, a protein synthesis inhibitor
Chl	chlorophyll
Chld	chlorophyllide
CK	cytokinin
CPM	counts per minute
DCIP/DCIPH$_2$	dichlorophenol-indophenol (ox/red)
DCMU	dichlorophenyl-dimethylurea, an inhibitor of electron transport
DMSO	dimethylsulfoxide, a solvent
DNA	deoxyribonucleic acid
DNP	dinitrophenol, an uncoupler
DPM	disintegrations per minute
EDTA	sodium (tetra) ethylene-diamine tetraacetate
EGTA	ethylene glycol bis (β-aminoethyl ether) tetraacetic acid
FID	flame ionization detector
GA	gibberellic acid
GC	gas chromatograph
HEPES	hydroxyethylpiperazine-ethane-sulfonic acid, a buffer

IAA	indoleacetic acid; auxin
MOPS	morpholino propanesulfonic acid, a buffer
NAA	naphthaleneacetic acid, an analog of auxin
NAD$^+$/NADH	nicotinamide-adenine dinucleotide (ox/red)
NADP$^+$/NADPH	nicotinamide-adenine dinucleotide phosphate (ox/red)
NAD(P)H	either NADH or NADPH
Pchld	protochlorophyllide
PGA	phosphoglyceric acid
Pi	inorganic phosphate
PPi	pyrophosphate
PMS	phenazine methosulphate
ppm	parts per million
P_r/P_{fr}	phytochrome
Ri-5-P	ribose-5-phosphate
Rubisco	ribulose bisphosphate carboxylase oxygenase
Ru-5-P	ribulose-5-phosphate
RUBP	ribulose bisphosphate
SCHB	a buffer for isolating thylakoid membranes, containing sucrose, choline and HEPES
SDS	sodium lauryl sulfate, a detergent
SGII	Sager and Granick medium
TCA	trichloroacetic acid
TIBA	triiodobenzoic acid, an inhibitor of the polar movement of auxin
TLC	thin layer chromatography